The Story
of the
COAST GUARD

1790–1939

Riley Shepard Brown

SUNBURY PRESS

Mechanicsburg, PA USA

Published by Sunbury Press, Inc.
Mechanicsburg, PA USA

SUNBURY
P R E S S

www.sunburypress.com

For information about special discounts for bulk purchases, please contact Sunbury Press Orders Dept. at (855) 338-8359 or orders@sunburypress.com.

To request one of our authors for speaking engagements or book signings, please contact Sunbury Press Publicity Dept. at publicity@sunburypress.com.

FIRST SUNBURY PRESS EDITION: February 2023

Set in Adobe Garamond Pro | Interior design by Crystal Devine | Cover by Lawrence Knorr | Edited by Sarah Peachey | Interior photos courtesy of U.S. Coast Guard.

Publisher's Cataloging-in-Publication Data
Names: Brown, Riley Shepard, author.
Title: The story of the Coast Guard : 1790–1939 / Riley Shepard Brown.
Description: First trade paperback edition. | Mechanicsburg, PA : Sunbury Press, 2023.
Summary: Riley Shepard Brown has recounted the history of the United States Coast Guard, from its inception in 1790 until the outset of World War I.
Identifiers: ISBN : 979-8-88819-075-3 (softcover) | ISBN : 979-8-88819-076-0 (ePub).
Subjects: HISTORY / Maritime History & Piracy | HISTORY / Military / Naval | HISTORY / Military / United States.

Product of the United States of America
0 1 1 2 3 5 8 13 21 34 55

Continue the Enlightenment!

Cover art: *U.S. Revenue Cutter, the U.S. "Morris"* painted after 1855 by H. A. Roath in the collection of the Philadelphia Museum of Art, courtesy of The Collection of Edgar William and Bernice Chrysler Garbisch, 1967.

To my shipmates, the officers and enlisted men of the Coast Guard—and to the memory of those men I knew; and who were killed in the line of duty; to the men without whose untiring devotion to duty, the stories recorded within would not have been possible, I dedicate this book.

—RILEY BROWN

The United States Coast Guard, one of the oldest agencies of the Government, exemplifies the history and time-honored traditions of our Nation. Down through the years, the Coast Guard has carried on its assigned duties thoroughly and efficiently. New problems are daily being created, added responsibilities are being given to the Coast Guard in the consolidation of Government agencies; all this tends to stress the importance of the Service to the average American.

The book written by one who is serving in the Coast Guard gives a keen insight into a part of the work that the Service is doing day in and day out. The past history of the Coast Guard is dwelled upon, as are the present-day aims and duties.

R. R. WAESCHE,
REAR ADMIRAL, U. S. COAST GUARD, COMMANDANT

Contents

ILLUSTRATIONS

FOREWORD

Being a military and naval historian, I often attend events honoring the armed services. At many of these, a band will play the "Armed Forces Medley," which, I have since learned, can be altered to put a particular service first or last. But for some reason, the segment with "Semper Paratus," the United States Coast Guard Anthem, is either left for last or, worse, not played at all. Even the shouted "Army, Navy, Air Force, Marines," rarely includes the Coast Guard. This is not only thoughtless; it is an insult to one of the five established armed services of the United States. But most people seem to consider the Coast Guard little more than a bunch of wanna-be sailors that rescue boaters or catch drug smugglers. At most, they accept that the USCG has done great work during natural disasters like hurricanes and shipwrecks.

When Sunbury Press publisher Lawrence Knorr asked me to write the Foreword to Riley Brown's book on the Coast Guard, I was not only flattered but enthusiastic. I was honored to be asked, and a bit ashamed. In the more than two hundred articles I have written for several national military history magazines, not one has concerned the fleet of beautiful white ships and boats that watch protectively along our coastlines. The United States Coast Guard is a true armed service, a close parallel to the Navy. But unlike the big gray ships that patrol the oceans keeping the peace, the USCG wears many hats.

To make the point that the Coast Guard does not just replace bulbs in lighthouses, there were USCG cutters and patrol craft at Pearl Harbor on December 7, 1941. Their crews fought back against the Japanese bombers, alongside their Navy brethren. A coast guard cutter was on station near Howland Island in July 1937 to aid aviatrix Amelia Earhart in her ill-fated flight around the world. The Coast Guard was in the First

World War, patrolling off the Eastern Seaboard. It was and still is a major element of the International Ice Patrol that charts and tracks icebergs in the North Atlantic to prevent tragedies like the 1912 sinking of *Titanic*. They were on hand to rescue survivors of the disastrous *Morro Castle* fire in 1934. Their cutters served off the coast of Vietnam during the war and were the first vessels to patrol the Mekong Delta in what would be called the "Brown Water Navy."

Wherever humans have gone to sea, the United States Coast Guard, true to their motto, "Semper Paratus," Always Prepared, is there.

Today the USCG has a new mission, interdicting drug smugglings and terrorist operations. They have saved countless lives from coast to coast, in hurricanes and fires, floods and wars.

This book was first published in 1939 and has long since been recognized as one of the best books concerning the history of the Coast Guard. While he does not cover the Second World War and beyond, Riley Brown wrote a highly informative and even entertaining account of the Coast Guard's origins, changes, and work through the turbulent years of the Great War and Prohibition. It is well worth reading. At last, with this Sunbury Press edition, the United States Coast Guard can be given the respect it has long lacked in American maritime history.

MARK CARLSON
Military Writers Society of America
San Marcos, California

Chapter I

FROM THE REVENUE MARINE
TO THE COAST GUARD

July 1789. A new nation had been created. Formed from thirteen states bonded together at first in the defense of common ideals, the United States were faced now with problems issuing from the peace for which they had fought so valiantly. They had to put into practical operation governmental departments strange and untried in the congress of nations.

So it was that on July 31, 1789, the president approved an act providing for the imposition of duties on the tonnage of vessels and on the importation of goods and merchandise into the United States, and established, in the spirit of the act, the need for a Revenue Marine, In September of the same year, Congress established the Treasury Department, and Alexander Hamilton was named as the first secretary.

As the Continental Navy had been disbanded at the close of the new war, there were no means of enforcing the newly enacted revenue laws until, at Mr. Hamilton's suggestion and with the president's full approval, Congress passed an act calling for the construction of ten vessels, the commanders of which were to operate directly under the Collectors of Customs in the enforcement of revenue laws. This, actually, was the beginning of the Coast Guard.

These vessels, designed for speed and ease of handling, put to sea upon completion, and for more than six years constituted the nation's sole armed force afloat. They enforced the revenue laws, afforded protection from pirates to the inhabitants of isolated points along the coast, and scoured the sea for the snaky ships of these plunderers.

Known at this time as the Revenue Marine, an act of February 1863 designated the service as the United States Revenue Cutter Service; and from 1868 until 1915, acts of appropriations referred to the service in this manner. The act of January 28, 1915, however, combined the United States Revenue Cutter Service with that of the United States Life-Saving Service and provided that the name, United States Coast Guard, be used henceforth.

The Coast Guard forms a very definite part of the armed forces of the United States. The act of February 1799 authorized the president to direct the Revenue Marine to cooperate with the Navy in defense of the coastlines. Since that time, the Coast Guard has engaged actively in every war in which this country has been concerned; but it was the act of January 28, 1915, that provided for the merging of the Coast Guard with the Navy at the outbreak of hostilities.

To provide for the professional education of young men who are candidates for commissions in the Coast Guard, an academy is maintained at New London, Connecticut. Appointments to cadetships are made upon strictly competitive educational examinations and are open to those young men of the prescribed age, seventeen to twenty-two years, who have the necessary moral and physical qualifications. These examinations are held throughout the country from time to time, candidates making the highest average being selected for cadet appointment. There follow four years of military and comprehensive training at the academy in New London, four years of hard work, during which the cadet is fitted, physically, mentally, and morally for his duties as an officer of the United States Coast Guard.

American naval history is sprinkled with heroic deeds performed by Coast Guardsmen, down from the days of the Revenue Marine to the present time. They have always been fighting men, because the Coast Guard was conceived for the precise purpose of fighting to preserve the revenues of the United States. In wartime, the Coast Guard has formed an invaluable adjunct to the Navy and has done much to add to the glory of the American naval forces.

On October 13, 1814, the Revenue Cutter *Eagle*, under the command of Captain Frederick Lee, was cruising in the vicinity of Long

Island, searching for the barges and tenders of the British brig *Dispatch*, which were in the Sound annoying shipping and causing great concern among the inhabitants ashore. Captain Lee was directed not to encounter the brig herself, as she was too heavily armed, manning eighteen guns to the cutter's six, but merely to attack and destroy the British vessel's attendant craft.

The *Eagle* came upon one of these small boats in a fog not far off-shore. The American cutter immediately gave chase, firing several shots into the sloop, but did not succeed in overhauling her. Around nine o'clock, the fog suddenly lifted, and the brig herself was discovered by the *Eagle* coming down upon the chase from a starboard course.

Seeing at once that he could not hope to escape to sea, Captain Lee realized that his only chance was to drive his vessel upon the beach and, removing the guns, defend her from a vantage point ashore.

The *Eagle* went aground just under a high cliff, and the tremendous job of dragging her guns to the top of the precipice was begun. Arriving off the *Eagle*'s position just as this operation was completed, the enemy brig immediately began firing. The first shell took effect in the *Eagle*'s hull, tearing a gaping hole just above her waterline.

Captain Lee assembled his men and guns on the cliff and began a heated answer to the enemy's fire. The British vessel attempted to get in close so as to rake the position of the Americans, but the rapid and efficient handling of the four-pounders prevented this. At this time, it was discovered that the wadding for the guns was running low, and since it was feared that the British might put a landing party ashore and try to storm the American position, Captain Lee decided to ask for volunteers to go back aboard the *Eagle* and secure all clothing and material that might be used as wadding. This, in itself, would be a risky undertaking, the captain pointed out, as the cutter lay squarely under the guns of the British warship and was subjected to continuous fire.

There was no hesitation on the part of the *Eagle*'s crew. As one man, the entire body stepped forward; and of these, five men were chosen and armed with muskets.

The five men started the perilous descent from the cliff to the beached cutter. The British trained their guns on them and tried to blast them out of existence. Shells burst all around them. A fragment struck the leader, wounding him severely. The remaining four men made him as comfortable as possible behind a pile of rocks and kept on.

Reaching the *Eagle*, they scrambled aboard in a hail of exploding shells to find the cutter a complete wreck. Gaping holes showed where direct hits had registered, and the masts were nothing but jagged stumps. As the flag had long since been shot away, one of the volunteers found another and hoisted it on the stern.

All available cloth and old log books were collected on the deck. The volunteers, knowing that the supply of bullets was growing scarce on the cliff, went around picking up the metal that had lodged in the *Eagle's* hull. Then they prepared for the dangerous trip back to the American guns.

The brig had shifted slightly and was now in a position to rake the open beach between the cutter and the cliff, rendering the situation extremely delicate for the four men on the *Eagle*. Captain Lee, as if sensing his men's circumstances, began a furious attack on the British warship, and under the cover of this fire, the volunteers dropped over the side and raced for the cliff.

Halfway across, a shell bursting directly in front of them instantly killed one man. Stopping only to pick up the material the man had been carrying, the little party kept going, finally reaching the rocks behind which they had left the wounded sailor. Almost immediately, the enemy ship was seen to get under way, moving closer in toward shore, and marines could be seen gathering at the rails. That could mean only one thing—a British landing party!

Carrying the precious wadding, the volunteers scrambled desperately up the cliff. They came upon Captain Lee standing on the edge of the precipice; he was studying the enemy vessel through a glass.

"Captain!" exclaimed one of the men. "They're coming ashore! With bayonets!"

Captain Lee smiled. "They'll never reach us," he replied.

They could hear the whistles on the British warship. As the sound of shouted commands came clearly across the water, Captain Lee tore the books and cloths into small pieces and distributed them among his gunners.

"Aim for their top-hamper, men," he said. "Make every shot count."

The brig lay close in, her white sails glistening in the sun, her decks a beehive of activity as the landing party prepared to go over the side.

Captain Lee raised his hand, hesitated, then brought it down with a snap.

The six guns roared as one. Instantly, bedlam struck the brig. Her sails were torn and her masts splintered by the shots. The whole aspect of the picture was changed. It became urgently necessary for the British to get their craft out of the range of the deadly little guns on the cliff, and this they did, working desperately while the Americans poured a hot fire down upon them. That night, while the enemy craft was disappearing into the distance, the cutter men stood by their guns and cheered.

This episode is only another of the many glorious incidents which are glowing pages in Coast Guard history. Big men in little ships, they have been in the thick of every fight, making traditions upon which the Coast Guard stands solidly today.

In the War with Mexico, 1845-47, eleven cutters, under the command of Captain Webster, took part in the hostilities. Arriving at Southwest Pass in the Gulf, Webster detached two of the cutters, *Forward* and *Ewing*, and directed them to report to General Taylor, commanding the American forces in Mexico.

At 8 P.M., on the evening of June 23, 1846, the two cutters came to anchor off the bar at Brazos Santiago, Mexico, and on the next day, Captain Nones, accompanied by Lieutenants Jones and Scott, left the vessel to communicate with General Taylor.

The general explained the situation to them, saying that it was urgently necessary for all supplies and munitions of war being shipped in to the enemy by way of the sea to be cut off and confiscated. Therefore, he ordered the two cutters to patrol the seacoast as far south as Soto La Marina and to capture and destroy all vessels discovered landing cargoes or in any way supplying the enemy with the means of war.

For three months, the two cutters in company with other vessels, patrolled the area assigned, effectively blockading the coast and doing the enemy incalculable damage. In October 1846, Captain Nones of the

Forward received orders from Captain Webster to proceed to the mouth of the River Alvarado, where in company with other units of the fleet an attempt would be made to enter the strongly fortified river and proceed inland.

This news was received with high enthusiasm by the men on the cutters for it afforded them a chance to break up the monotonous grind of patrol duty. A little after sunrise on the morning of October 15, the *Forward* reached the mouth of the river, where she found already assembled the steamer *Vixen*, three gunboats, the prize *Nonata*, the *Mississippi*, and the cutter *McLane*, all under the command of Commodore Perry in the *Mississippi*. At a council of war held aboard the Cutter vessel, Captain Nones learned that the army wanted the destruction of enemy supply bases and munitions dumps up the river. Forts on both banks commanded the approach from the bar, making the task a difficult assignment.

It was decided that the *Mississippi*, having the heaviest guns, should move in and cannonade the forts while the smaller ships crossed the bar. Unfortunately, it was found in actual operations that the *Mississippi* could not approach close enough to make an impression on the forts with her shells. These forts mounted a battery each of seven guns, and directly behind the forts were two powerful pivot guns capable of throwing a shell down into the midst of the American fleet.

Despite all obstacles, Captain Nones decided to make the attempt to enter the passage. He signaled his intentions to the *Mississippi* and received an approval, together with the suggestion that he wait for a favorable breeze which would help the smaller vessels. Captain Nones delayed his attack until two o'clock, at which time—there being no prospect of a favorable wind—he decided to move at once. Dividing his force into two sections, with the smaller vessels in tow of the larger ones, the captain started his advance.

The first division, consisting of the *Reefer* and the *Bonita*, in tow of the *Vixen*, came upon the bar and at once engaged the enemy. The second division—the *Nonata*, *Forward*, and *Petrel*, towed by the *McLane*—followed. The first division was well within the bar and was matching fire with the forts when misfortune fell upon the American fleet. A shell from the shore batteries struck the *McLane*, damaging the steering apparatus slightly and throwing her aground on the bar, with the tow, the *Nonata*, *Forward*, and *Petrel*, afoul of each other in wild disorder. Seeing his advantage, the enemy immediately began a terrific fire upon the fouled ships.

Miraculously, the division suffered no critical injury. Captain Nones, dropping his tow, put about in the *Vixen* and went to the assistance of the *McLane*. With the combined efforts of the two vessels, the *McLane* was floated and the tow reorganized. The tide had begun to run out, however, and Captain Nones decided to postpone the attack, for he felt that the *McLane*'s draft was too deep to allow her to enter at low water. Reluctantly he signaled his flotilla to turn and withdraw.

With the intention of catching the enemy unaware, the next attempt by the Revenue Cutters to enter the river was made under cover of darkness on the morning of October 23. The plan was suddenly discarded, however, when it was found that an American trading bark, with supplies for the enemy, had been in treasonable communication with the latter and had signaled the forts of the Americans' advance. This bark was taken into custody by the *Vixen*, after which the American fleet sailed across the bar, and at once engaged the Mexican forts.

Shells framed through the darkness. The *Vixen* was hit twice, the *Forward* once. The great guns of the Mississippi threw shell after shell into the forts. Presently, the fire from the enemy slackened. Finally it stopped altogether.

By this time, the American fleet was beyond the bend in the river, one objective reached. The main task, however, lay ahead; namely, the destruction of the military warehouses at Tabasco, a strongly fortified city 75 miles upstream. Accordingly, the fleet pushed on, and in the evening of the next day anchored in battle formation, half-musket range, before the city.

Commodore Perry sent a dispatch ashore, ordering the city to surrender immediately or suffer the consequences of a bombardment. The Mexicans returned the dispatch, with the invitation to begin firing at will, as the city did not intend to surrender.

Accordingly, the *Forward*, was directed to open fire, first on the flag staff over the city. This was done, the first few shots carrying away the flag. As the flag was not replaced, the American fire was ordered held, and a deputation was sent ashore to learn if the colors had been struck in token of surrender. The answer was the same—no surrender.

Commodore Perry ordered the *Vixen* and the *Forward* closer inshore and directed the *Vixen* to land a force and take up a position in the city, which was commanded by the guns of the fleet. This operation was disputed hotly by musketry from houses along the shore, and the Revenue Cutters poured round and grape shot into the buildings in an effort to silence the guns. After twenty minutes of this, no further opposition was encountered by the landing party, and they secured the waterfront, burning the warehouses which obviously held war supplies.

Toward evening, Commodore Perry ordered the landing force back aboard the *Vixen* and sent word to the Mexican authorities that if the city was not surrendered by morning, he would endeavor to destroy it. There was no answer.

In the early hours of the morning, a furious fire was begun by the batteries on shore, and the advance position of the Revenue Cutters became extremely hazardous. But they did not fall back; instead, the Revenue Cutters used the short distance between them and the enemy guns to great advantage, spraying the waterfront with grape shot. After an hour of continuous bombardment by the American ships, a flag of truce was displayed over the city, and Captain Forest, commanding the American Marines, was sent to converse with the Mexicans. The city authorities asked for a suspension of hostilities for a number of hours until the wounded could be removed from the scene of action. This request was granted, but in the light of consequent events, it proved nothing more than a ruse to gain additional time.

Commodore Perry now ordered the prizes which had been taken in the river to proceed to the mouth and await the flotilla there. In complying with these orders, one of the prizes drifted ashore in front of the city, and the enemy, collecting a large force with heavy guns, opened fire upon the ship in violation of the truce. Lieutenant Parker of the Revenue Marine was in command of the ship. He defended her gallantly, ultimately succeeding in working the vessel free, but only after a number of the Revenue men had been killed and wounded. The guns of the flotilla were again turned on the city. A landing force from the *Forward*, commanded by Lieutenant McGowan, went ashore and routed the enemy concentration while grape and round shot from the *Forward's* gun fired the waterfront.

The entire city was now in flames, and all organized opposition had ceased. Commodore Perry ordered the landing force back aboard the *Forward*. Then the fleet dropped down the river, its objectives gained.

In schoolrooms all over the country may be found a copy of the famous picture showing the Revenue Cutter *Hudson* taking up the tow of a disabled sister ship, the *Winslow*, during an engagement with the Spanish land batteries at Cardenas Bay, Cuba, on May 11, 1898, in the Spanish-American War. This picture shows shells bursting on and around the *Winslow* while the Revenue Cutter *Hudson* dashes in to pull the disabled ship to safety.

The picture is an inspiring one, typically American, typically Coast Guard. Here, in the heat of battle, the duty which is primarily the essence of the Coast Guard came to light. Lives had to be saved, and, discounting all risks, the Revenue Cutter went about doing that duty.

On the morning of May 11, 1898, the Hudson, under the command of Lieutenant Frank H. Newcomb and accompanied by the USS *Wilmington* and the U.S. torpedo boat *Winslow*, steamed into Cardenas Bay, Cuba, on the lookout for enemy ships. Once inside the bay, the senior officer present on the Wilmington detached the *Hudson* and sent her on patrol duty along the western shore of the bay.

A Spanish gunboat came into view around the curve of the island, stopped, and waited, evidently being aware that the shallow water would keep the Revenue Cutter at a distance. For some time, the *Hudson* steamed back and forth in an effort to draw the enemy ship out to battle, but failing in this, observed that the *Wilmington* and the *Winslow* were nearing Cardenas, with signals flying directing the *Hudson's* return.

The Revenue Cutter immediately turned and headed full speed toward the two naval ships, and when a mile distant, saw smoke rising from the middle of the waterfront, and the report of a heavy gun rolled across the bay. Instantly, the *Winslow* turned and headed in toward the wharves, with the *Wilmington* taking up a battle position further out. The American ships answered the shore batteries, and by the time the Revenue Cutter arrived on the scene, about 1:45 P.M., a general engagement was in progress. Lieutenant Newcomb brought the *Hudson* in between a bark and a brig lying at anchor and, taking a position about 1,800 yards offshore, off the western end of the city, opened fire with her two six-pounder Driggs-Shroeder guns upon the enemy battery, which was located in the center of the waterfront.

Meanwhile, the *Wilmington* had turned and was steaming slowly to the westward, outside of the *Hudson's* position. The *Winslow* still maintained her original station, which was several hundred yards inshore from the *Hudson*. After firing a few rounds from the *Hudson*, Lieutenant Newcomb, noticing that the tide was swinging his vessel around into the line of fire from the *Wilmington*, ran around the naval vessel to obtain a clearer field for operations. Passing within hailing range of the *Wilmington*, Newcomb pointed toward the Winslow and asked, "Shall I take up position there?" The answer came back, "Yes."

The *Hudson* ran in at full speed until about 150 yards inshore of the *Winslow* and a short distance to the eastward of her. The former's engines were stopped, and firing was resumed with the two six-pounders. Observing the effect of his ship's fire through his glass, Lieutenant Newcomb saw that the shells were falling short of their mark and directed that the elevation be raised. After this, the *Hudson's* shells appeared to be falling within the enemy's works, which could be distinguished beneath a hovering pall of smoke.

At this time, the *Winslow* abruptly got underway, moving in such an aimless fashion that a collision with the torpedo boat was narrowly avoided by Lieutenant Newcomb. The *Winslow* kept running back and forth on a parallel with the shore, smoke rising from her in great sheets. Thinking that her ammunition had exploded, Lieutenant Newcomb signaled the torpedo boat and asked if she was in need of assistance. The answer came back that she had been hit but that no assistance was needed.

The enemy shells were falling and bursting all around both vessels. One large shell passing closely over the top of the *Hudson's* pilot house struck the former vessel but failed to explode. Apparently the enemy was bringing up heavier guns now, and fire seemed to be concentrated on the *Winslow*, which was still maneuvering wildly to the eastward of the *Hudson*. Several direct hits were registered on the torpedo vessel, and it became apparent to Lieutenant Newcomb that she was badly hurt and now needed assistance.

Again, he signaled an offer of aid, and this time the *Winslow* reported herself totally disabled and requested a tow out of the range of the enemy guns. In starting to her assistance, Lieutenant Newcomb was compelled to bring his vessel directly in the line of fire from the *Wilmington*, some of whose shells, exploding soon after leaving the guns, sprayed the Revenue Cutter with fragments.

A stiff breeze blowing obliquely on shore from the eastward caused the *Winslow* to make much leeway, and as she was constantly shoaling the water, Newcomb found it extremely difficult to throw a line to her. The water was so shallow that the *Hudson*'s propeller was constantly throwing up mud, and steerageway was invariably lost as soon as the speed slackened.

As the *Hudson* approached the disabled vessel, an officer, Ensign Bagley, and three enlisted men were standing on the *Winslow*'s bow waiting to take the Revenue Cutter's line when a large shell exploded directly in their midst, killing them instantly. Obviously trying to prevent the Revenue Cutter from drawing the disabled vessel out of the range of the shore battery, the enemy directed an intense fire at the two vessels.

During these operations, the *Hudson* was struck at least four times, but none of these projectiles did any great damage. Attempts were made to get a line over to the *Winslow*—which was a complete wreck—but these efforts were being defeated by the shallow water into which the two vessels had worked themselves. It appeared that both would go aground and thus be at the mercy of the Spanish guns.

But these were difficulties under which the Revenue Cutter men excelled. Lieutenant Newcomb brought the *Hudson* around and, edging past the *Winslow*'s bow, finally managed to get a line over to her. The *Hudson* took up the strain, and the struggle toward safe water was begun.

The *Winslow*'s steering gear had been wrecked, and she yawed widely, threatening to foul the Revenue Cutter with her hawser. With consummate skill, however, Newcomb brought the disabled vessel out of range of the shore batteries.

Meanwhile, the *Wilmington* had silenced the enemy guns, and, apparently unaware of the *Winslow*'s critical condition, she was standing down the bay. Both the *Winslow* and the *Hudson* tried to signal her for a doctor, but the flagship was a mile away now, and the signals went unanswered.

The *Hudson* and her tow had reached deep water when, due to the yawing of the *Winslow* and the choppy sea, the hawser parted, and the torpedo boat wallowed helplessly. Another hawser was passed to her, and as a temporary rudder had been rigged on her stern, the tow was resumed, this time with greater success.

After a long and laborious chase dead to the windward, the *Hudson* finally overtook the *Wilmington* and, hailing her, requested that a doctor be sent aboard the *Winslow*. After the latter's dead and wounded had been removed to the flagship, the Revenue Cutter made fast alongside the *Winslow* and attempted to tow her in that manner. However, as the water thus thrown up between the two vessels threatened to sink the disabled craft, the original method of towing was resumed. Steaming out of the bay, they finally reached the U.S. supply ship *Machias* at nightfall.

In his report to the Navy Department, Lieutenant Bernadou, who was in command of the *Winslow*, said, "—without the combined efforts of all those on board the *Hudson*, the surviving members of the crew of the *Winslow* would have been lost."

Chapter II

IRON MEN, WOODEN SHIPS

A year after the thrilling work of the Coast Guardsmen aboard the *Hudson*, another little band of these heroes wrote heavily into the book that records some of the thrilling sagas of the sea. In August 1899, a vicious hurricane swept northward along the coast, and today there remain jagged timbers of the various ships which were wrecked by this catastrophe. At low tide, the hulk of the three-master schooner *Aaron Reppard* may still be seen on Hatteras Island, North Carolina, a few miles below the Gull Shoal Coast Guard Station. About a week before the wreck of the *Reppard*, this hurricane had swept over the island of Puerto Rico, wreaking untold damage and killing hundreds of persons. In the intervening time, it had slowly moved northward, slightly diminished in intensity yet still maintaining its tremendous destructive force. On August 13, when the storm's center was off Jupiter Inlet, Florida, shipping interests had already been advised of its approach up the coast, and vessels were warned to seek shelter.

It is not known whether Captain Wessel of the *Reppard* was actually aware of impending danger. He had left Philadelphia on the morning of August 2, and was now, at this time, the fourteenth, off the capes of Delaware, standing south with an easterly wind for Savannah. By 8 P.M., the increasing easterly wind had made necessary the taking in of all light sails and the setting up of preventer stays.

The trim little vessel rode easily through the night, and by the morning of the fifteenth, while the hurricane was raging around the port of her destination, she was somewhere off Cape Henry.

The signs of the advancing storm were fully discernible now to the *Reppard*'s crew. The sky was of a heavy slate color, and seagulls were flying swiftly northward. The glass in the captain's cabin was low, indicating that a disturbance of a violent nature was at hand. Several of the crew went to Captain Wessel and asked if he intended on taking the schooner into the safety of nearby Chesapeake Bay. Wessel's replies are not known, but later events proved that he held to his course until 4 P.M., when he was compelled to heave his vessel to and try as best he could to ride out the storm, with all possible chances of reaching calm waters gone. Two hours later, the wind reached tremendous force, blowing so hard, in fact, that the tops of the waves were blown completely away, and the sea became comparatively calm.

But this was not to last. The center of the storm shifted slightly; the sea started making up again, building towering waves that roared down upon the luckless schooner. She became badly strained in fighting these seas, and the crew was ordered to the pumps. All that night, the *Reppard* remained hove to on the starboard tack under fore staysail and reefed mainsail, with her helm lashed hard down. Giant combers swept the decks, beating her bow down, and on Wednesday morning, the mizzen storm trysail was set to hold her up.

The weather had become very thick mist from the sea and a heavy rain mingling to reduce visibility to zero. Realizing that his vessel was being constantly driven shoreward, Captain Wessel tried every way possible to haul her away from this danger, for he had reason to believe that the *Reppard* was not far off the beach and would strike before long.

At 1 P.M., when breakers were sighted astern, Wessel immediately ordered the staysail to be taken in and both anchors let go in an effort to stop the *Reppard*'s drag toward destruction. Ninety fathoms of chain were let out, and the ship shivered and staggered as the anchors dug into the bottom. But it was of no avail. The anchors could not hold her against the tremendous seas. Within a short time, she was in the first line of breakers, with grounding imminent.

The crew now let go the mainsail halyards, and all hands were ordered into the shrouds to escape the gigantic breakers that were sweeping the decks. The crew numbered seven men, and there was one passenger—a Mr. Cummings, from Charleston, South Carolina.

The *Reppard* had long since been sighted by the surfmen on the beach. Surfman William G. Midgett, who was on beach patrol at this time and the first to sight her, wrote in his official report: "I sighted her masts through the murk for the first time when she was about a mile and a half off the beach. The schooner was heading north, now making a little headway, then dropping back. She came into the breakers in a few minutes, and I left immediately to notify my station."

The distance Midgett had to travel was about two miles, and although he was mounted and rode his horse hard, the weather conditions were such as to require twenty-five minutes for him to make the trip. As later events proved, however, he was in ample time to affect a rescue. Keeper Pugh, in charge of station (who, by the way, would have been known by the title of boatswain in these days), immediately telephoned Little Kinnakeet Station, the nearest southern station; Chicamacomico Station, the nearest northern station to Gull Shoal; and requested the keepers to join him with their crews abreast of the wreck.

Then Keeper Pugh attached horses to the beach carts, loaded lifesaving equipment, and set out for the wreck. Within ten minutes of his arrival, the crews from the other surf stations arrived, and preparations were at once begun to bring ashore the *Reppard*'s crew.

Keeper Pugh says in his report of the rescue operations: "The schooner was by this time about 700 yards offshore, stern toward the beach, riding to two anchors, but slowly dragging shoreward. This portion of the beach consists of two banks about 50 yards apart with a gulley between them, and the seas, which were as high as they could be, were sweeping completely over the land from the ocean side of the sound. It was my opinion, which was concurred in by Keepers Hooper and Midgett, that the use of a boat in these conditions was clearly beyond all realm of possibility. No number of men, however skillful, could have launched a boat in those seas."

As the wind was blowing in from the beach, chances of getting a line over to the *Reppard* were slim indeed. The Lyle line throwing gun was set up, however, and several unsuccessful attempts to get a running line across to the doomed vessel were made. Again, when the *Reppard* was some 500 yards off the beach, another attempt was made to fire a line across to her. This time, the line carried almost to her and then parted from the force of the shot. Keeper Pugh ordered a second charge set up, but with a heavier line, and these precautions resulted in success. The line went sailing into the shrouds of the schooner, to which a number of men could be seen clinging.

No attempt was made by the distressed men to reach the line. Van der Graaf, one of the survivors, said that it was fired perfectly and came directly across the ship, but that it was impossible for any of them to reach it. Indeed, it was all they could do to hold on.

It was evident now that the wreck was about to go to pieces, and the only thing that the lifesavers could hope to accomplish was to rescue the survivors from the surf as they came in. The *Reppard* was breaking up. The deck house went first, then the hatch coamings and the bulwarks. While this was taking place, Cummings, the passenger, aloft in the mizzen shrouds, was caught by one leg in the ratlines and slammed back and forth against the mast until his life was beaten away. Suddenly, the mast fell, and Cummings was not seen again.

The mainmast went next, breaking into two pieces, throwing the seaman Tony Nilsen into the debris. Badly hurt, the man screamed in pain as he lay in the wreckage. Then, some measure of strength returning to him, he crawled to the side, dropped into the sea, and was washed away.

Captain Wessel was the next victim. Just before the mainmast fell, he leaped into the sea and made a brave effort to reach shore. The men in the shrouds watched as he fought his way shoreward, now making a little progress, now falling back, sorely baffled by the whip-like lash of the seas, until it was apparent that his strength was failing. Desperately, he turned back toward the ship, his face twisted and strained with the agony of his efforts. His struggles carried him within five feet of the *Reppard*'s side. The men aboard her were shouting encouragement when, suddenly, a peaceful look came over his face; he closed his eyes, raised his arms above his head, and sank quickly from sight.

Four men remained alive aboard the *Reppard*—the mate Stewart Robinson, seamen Pedro Lachs, James M. Lynott, and Van der Graaf. These men knew that it was only a question of minutes before they, too, would be hurled into the raging sea. They were in the foremast rigging, and the foremast itself was straining and weaving with the motion of the wrecked schooner. Suddenly, the mast snapped off, and the men went hurtling down with it. Lynott fell upon the stump and was impaled upon it. For a few minutes, his screams rose above the howl of the wind, then, mercifully, a boiling comber washed him into the sea.

Fortunately, the mast had fallen over that side of the schooner, which was nearest to the shore, and the three men in the water had a bare fighting chance for their lives. A number of the surfmen at once donned cork jackets, and each taking about 50 yards of shot line, waded as far as possible out into the surf while each line was held by two surfmen on the beach.

These operations were attended by great peril to the surfmen in the water, for pieces of wreckage from the schooner were constantly being driven ashore with express-like speed so that the men had to move with great skill and rapidity to avoid them. The veteran keeper of the Little Kinnakeet Station, ignoring the advice of his brother officers to leave the hazardous work to younger men, was struck by a heavy timber while rushing in to steady a reeling surfman. He was pulled from the surf almost drowned, with fractured bones in his leg.

Clinging to wreckage, the three men from the *Reppard* were being gradually driven into the surf. One of them, too weak to hold on any longer, slipped away in the rolling breakers. A surfman quickly tied a line around his own waist and managed to get a hold on the drowning man. Both were pulled ashore, more dead than alive.

By dint of diligent effort, all three of the shipwrecked men who escaped alive were rescued from the surf. Too weak to stand, they were taken in beach carts to Gull Shoal Station and were given stimulants, wrapped in blankets, and placed in bed. One of the rescued men, Van der Graaf, suffered pneumonia from his terrible experience but recovered to spend many more years at sea.

The names of the rescued men were Bernard Johnson, Pedro Lachs, and John Van der Graaf; the five who died were James M. Lynott, Tony Nilsen, W. Robinson, Oscar Wessel, and the passenger Cummings. Only one body was recovered, that of W. Robinson, which was buried on the beach near Gull Shoal Station.

Lieutenant C. E. Johnston, who investigated the circumstances of the wreck and subsequent rescue for the Revenue Cutter Service—as the Coast Guard was known in those days—said in his report to headquarters:

"There is no doubt that the surfmen did everything humanly possible under the adverse conditions to save the lives of the people on the schooner. The storm was the worst in the recollection of any one now living on the Carolina coast and it is little short of a miracle that any one now lives to tell the tale of the wreck. If the master had not anchored, or if he had slipped his cables as soon as he reached the breakers, it is probable that all hands would have been saved, as the schooner would not have stopped until she was right up against the bank. Three other schooners, a barkentine, and a lightship all went ashore in the same general locality and in the same storm without anchoring, and the only loss of life from the five vessels was occasioned by a tremendous sea which boarded the barkentine when she first took bottom and washed four persons overboard. All the rest were rescued by the lifesavers."

Perhaps one of the most unique rescues in the history of the Coast Guard occurred when the barkentine *Priscilla*, bound from Baltimore to Rio de Janeiro, was blown ashore and broken up by the same hurricane that wrecked the *Reppard*. As a matter of peculiar coincidence, the *Priscilla* went ashore only a few hundred yards away from the dismal remnants of the *Reppard*, following the latter vessel to destruction but a few hours after the rescue operations just described.

The hurricane did not reach its full violence until August 17, at which time the Weather Bureau recorded a high of 120 mph. The *Priscilla* had fourteen persons aboard, twelve of whom comprised the crew, officers, and men. The remaining two persons were the captain's wife, Virginia, and their twelve-year-old son, Elmer.

On the morning of Wednesday, the sixteenth, the *Priscilla* was somewhere off the Virginia coast. The weather was clear, with a hard wind blowing from the southeast. Light sails were taken in, and by midday, it was found necessary to furl the spanker and upper topsail. With the wind still increasing, the foresail was hauled up and furled, the lower topsail, mainsail, and main staysail having blown away.

Commanded by the veteran Captain Benjamin E. Springsteen, the *Priscilla* was now hove to under bare poles and being rapidly driven south-southwest. Although Captain Springsteen had been unable to take any observation within the previous twenty-four hours, he knew that he was only a little to the northward of Cape Hatteras—which jutted out into the ocean in myriad finger-like shoals, a graveyard of many a ship. He did everything in his power to bring the vessel clear of these reefs, although, at the time, he must have entertained little hope of accomplishing any measure of success.

Around ten o'clock on the seventeenth, one of the officers reported that the water about the ship was discolored, indicating that they had been driven into the Gulf Stream. The master immediately ordered soundings to be taken, and when only "twenty fathoms" was reported, he realized beyond doubt that the vessel was rapidly being driven ashore.

At regular intervals, soundings were taken, and at each sounding, the depth of water decreased. At 8 P.M., when ten fathoms were reported, Springsteen ordered the soundings to be discontinued and directed that "all hands prepare to save themselves."

At ten minutes past nine, the time when the Weather Bureau reported the hurricane to be at its highest velocity, the *Priscilla* struck ground lightly but shipping enough water to smash the cabin skylights and deluge all below. The vessel did not strike again for about twenty minutes, at which time the impact was so terrific as to carry away the foremast and make of the deck a shambles.

Captain Springsteen immediately ordered all hands to the topside as the vessel was pounding heavily and rapidly becoming filled with water. Breakers sweeping over her from bow to stern made it difficult for the people on deck to hold on.

The ship's boy, Fitzhugh Lee Goldsborough; the cook, Leon Navens; the first mate, and Mrs. Springsteen were holding onto the deck house. Mrs. Springsteen was singing and comforting her small son, Elmer, whom the captain was holding. The ship gave a sudden lunge; the deckhouse dissolved into a mass of whirling green water and flying wreckage, the four persons holding onto it being swept overboard. Not one of them was ever seen again.

At the same time, Elmer Springsteen was torn from his father's arms and literally thrown into the sea. By some strange quirk of fate, his body was found later in the water-filled cabin, having been washed back aboard the *Priscilla* in some unaccountable fashion.

A short time later, the ship broke entirely in two, and each section continued to rake and pound along the beach in the breakers. Fortunately, the survivors were crowded upon the larger of the two sections, and they managed to lash themselves together to keep from being washed away. For five horrible hours, this nightmare of torture continued, with the wrecked hull disintegrating bit by bit. The weather was so thick as to make visibility zero, thus rendering any possible means of signaling to the shore ineffective.

At three o'clock that afternoon, Surfman Rasmus S. Midgett, of the Gull Shoal Station, was patrolling this section of the beach. The wind was blowing so hard that he found it hard to stay on his horse. Finally, dismounting, he placed cloth around the animal's head to protect it from the driving sand. At this time, he discovered various objects—buckets, barrels, boxes, and other articles—coming ashore in the surf. All this satisfied him that a vessel was breaking up somewhere in the vicinity and that it was up to him to find it.

The surf was sweeping clear across the narrow strip of land which separates the ocean from Pamlico Sound, and the water at times reached his saddle stirrups. The day was so dark that Midgett could barely see, but he drove his horse on, finding more and more wreckage along the beach.

Finally, about two miles from the point where he had first seen signs of the shipwreck, he thought he heard cries over the pound of the surf. Stopping his horse, he dismounted and, protecting his head against the animal's side, listened carefully. Yes, he was right. Cries for help came weakly with the howl of the wind.

Midgett prepared at once to get back to his station for help but stopped when the pall suddenly lifted from the sea and a portion of a ship was visible a few hundred yards away. Men were crouching on the hulk, which would disappear under a spume of spray and green water one moment, the next appear again a little closer in.

Here, Midgett was faced with the problem of making a tremendous decision. He knew that it would take too long to return to his station to secure help; on the other hand, if he were to lose his life in a fruitless attempt to rescue them, the men on the doomed vessel would be left without anyone on the beach being aware of their critical situation. Here, then, was a moment that called for sound judgment and unerring courage.

Midgett spent little time in deliberation. The saving of lives in these moments depended upon him alone. He rose to the demand.

Waiting until a wave receded, the surfman turned and headed his horse toward the wreck, driving the animal through the surf as hard as he could. Before the sea came rushing in again, he made it down to the wreck, his horse swimming against the pull of the tide. He called instructions to the men crouched on the hulk, telling them to throw themselves, one by one, into the sea and that he would pull them out. Then he retreated to the beach to await an opportune moment to begin his rescue work.

On the wreck, the sailors were in a quandary. They knew that they couldn't remain aboard the hulk much longer. Nor could they bring themselves to trust the surfman. It seemed impossible, the thing that he had promised to do. Then Captain Springsteen made up his mind.

"That surfman means what he says," he declared to the rest of the survivors. "If he has the nerve to come down here to help us, I'll do as he says. Here goes!"

Captain Springsteen crawled to the rail and, waiting for the sea to recede, dropped over the side.

Instantly, Midgett pounded into the sea, drove his horse down close to the wreck, and pulled Springsteen out to the beach, reaching safety only a step or so ahead of a wall of roaring water.

Following their captain's example, all but three of the remaining sailors threw themselves, one by one, into the sea, and Midgett pulled them to safety. Each trip, the surfman risked the grave danger of being caught by the breakers and swept out to sea with his burden. But now, having rescued seven of the shipwrecked sailors, he faced even greater demands upon his courage and physical stamina. There remained aboard the wreck three men who were so badly bruised and exhausted that they were unable to throw themselves into the water as their companions had done.

Surfman Midgett was not dismayed, despite the fact that his horse was now exhausted and would prove of no further use to him in rescuing the three remaining men. Leaving the animal on the beach, he grasped the line that is a part of every surfman's equipment and plunged into the surf again.

It was a hard battle, one that taxed Midgett's superb physical prowess to the limit, but he reached the wreck and managed to clamber aboard. With his line, he lowered one of the injured sailors to the water; then, selecting an opportunity when the breakers had receded, he took a hold on the injured sailor and struck out for safety.

Disaster almost caught them. The surf came back just before they reached the beach and tumbled them over and over in the sand. Captain Springsteen, with two of his men, ran out and helped them ashore.

But the surfman's task was not finished. Exhausted though he was, he knew that as long as there remained a single person aboard the wreck, he must carry on. Resting for a few minutes, he regained a measure of his strength and twice more went aboard the wreck. Each time, after laborious exertions, he brought a man to safety.

Ten lives were in this manner saved. Three of these sailors were so grievously injured, a physician determined later, that they could not have withstood the exposure for another hour. The seven men who were able to walk Midgett sent toward the station and, taking the horse, rode on to summon help.

He met his superior officer, Keeper Pugh, on the beach in front of the station, and the veteran officer could scarcely believe the amazing story that Midgett related. Beach carts were sent to bring the shipwrecked men into the station, where warm clothing and a doctor awaited them.

Midgett's exploits in this day's work made headlines in many of the nation's papers. The Secretary of the Treasury, under whom the life-saving service operates, presented the surfman with a gold life-saving medal of honor, and transmitted with it a long and highly commendatory letter, describing in detail the story of this brave man's heroism.

THE COAST GUARD AS IT WAS—1898

Coast Guard Cutter *Hudson* rescuing the disabled torpedoed *Winslow* from under the batteries at Cardenas, Cuba, May 11, 1898 (from a painting).

MODERN COAST GUARD SPEED BOAT AND AMPHIBIAN ON A RESCUE

Speed boat and amphibian plane head out at full speed to effect a rescue off the coast.

THE CRUISING CUTTER *PONTCHARTRAIN*—BUILT 1928

Displacement, 1,979 tons; length, 250 feet; beam, 42 feet; draft, 16 feet; power, turbine electric; maximum speed, 16.5 knots; maximum cruising radius, 7,542 miles; guns, two 5-inch, two six-pounders; one 50-caliber anti-aircraft, one one-pounder; personnel, 13 officers, 98 men.

A 165-FOOT PATROL BOAT—BUILT 1933

Displacement, 334 tons; length, 165 feet; beam, 25 feet and 3 inches; draft, 8 feet and 6 inches; power, Diesel; maximum speed, 16.5 knots; maximum cruising radius, 6,000 miles; guns, one 3-inch, two one-pounders; personnel, 5 officers, 37 men.

Midgett's reaction to these honors was typical of the service. He said, "Anyone would have done what I did. It was my job. I don't deserve any medal."

During the night of January 20, 1903, the barkentine *Abiel Abbott*, bound from Turks Island, West Indies, to New York City, with a cargo of salt, encountered a sleet-laden storm. She carried a crew of nine men and was under the command of Captain Israel B. Hawkins.

At four o'clock that afternoon, Captain Hawkins had taken bearings on a lighthouse, which he supposed to be Barnegat light, off the New Jersey coast, but which was in all probability Absecon light. The captain set his course by this erroneous bearing, which, consequently, carried him ashore on the outer edge of Ship Bottom Bar.

At 8 P.M., the *Abbott* went broadside up on the bar and held fast. Sails were furled at once and signals of distress burned. Almost immediately, the red Coston signal of the beach patrolman, Surfman Pharo of Ship Bottom Station, answered this plea for assistance. Pharo started off at once to notify his station keeper; he arrived at the station a little after 9 P.M., when Keeper Truex immediately assembled his wreck crew, notified the neighboring life-saving stations (Harvey Cedars Station and Long Beach Station), and asked that their crews be sent to the scene of the wreck.

Meanwhile, the *Abbott* was resting comparatively easy, with her sails neatly furled, her men awaiting help from shore, but conditions were growing worse. The tide was making up, as well as the wind and sea, and the force of the waves soon drove the vessel farther on the bar, causing breakers to wash over the decks.

The life-saving crew from the Ship Bottom Station arrived on the scene and the Lyle gun was quickly set up. The first shot did not quite reach the *Abbott*, although her crew stated later that the projectile landed near them. Another charge was fired; this time, the shot landed aboard her. However, her men could not find the line for the reason that it fell amidships or forward, where the seas made it impossible for them to move to that portion of the ship.

The entire hull, with the single exception of the quarter deck, was completely submerged now, and the constantly increasing waves rolled with great force over it. A single glimmer of light denoted the position of the wreck, making it difficult for the surfmen on the beach to fire a line over it. Finally, when the men on the *Abbott* had failed to draw in any of the lines, these operations were discontinued with the disheartening knowledge that, although help had been sent across to the wreck, the crew had not, for some reason unknown at the time, availed themselves of it.

By this time, the crews from the adjacent stations had arrived, but there was nothing to be done in the darkness. The single light on the *Abbott* had become extinguished and the night was as black as ink. Only the faint whiteness of the breaking surf could be made out. No surfboat could be launched successfully under such conditions. The keepers decided that it would be best to wait until daylight before resuming rescue operations.

Accordingly, huge fires were lit on the beach in order to let the wrecked men know that they had not been abandoned. Back on the *Abbott*, those fires did a lot to hearten the stranded sailors, who thought that if the old hulk would only hold together until morning, they might all escape. Sheltering themselves as well as possible, they prepared to wait the night out, listening anxiously to the ominous creak of the timbers as tons of water pounded against the wreck.

Between three and four o'clock in the morning, the first sign of approaching disaster came. The mainmast fell with a loud crash and, held to the other two masts by stays, tugged and pulled until these remaining masts, weakened by the constant roll of the ship, fell at about five o'clock.

When the mainmast fell, seaman Timothy Brandt stripped to his waist and, despite the protests of his shipmates, jumped over the side with the intent of swimming ashore. His mind must have become unbalanced from the strain of the night's ordeal, for the survivors related that he paid no attention whatever to their advice but merely insisted over and over that someone wanted to see him on the beach. He never reached shore, and later his body was found at the surf's edge five miles below the wreck.

Eight men were still alive under the lee of the cabin but were subsequently washed overboard by a huge breaker. The masts were down in the water and were held to the ship by a wild tangle of stays and gear. Five of the eight men managed to climb back aboard by means of these lines and remained huddled on top of the cabin. Unable to get back to the masts, the mate and the cook were drowned.

One of the seamen, Henry Carter, succeeded in grasping the end of the spanker gaff, which hung in the water; evidently, he was injured by falling wreckage. He called weakly once or twice, begging his shipmates to throw him a line with which to secure himself. No one could aid, and he was soon swept away.

Five men remaining on top of the cabin—which was exhibiting signs of breaking up itself—prayed for daylight. Soon dawn came, a muddy gray light that revealed a dismal scene. On the beach, the surfmen heard faint cries and immediately jumped for their surfboat. Wreckage filled the breakers; launching was a hazardous job. Nevertheless, one crew manned the boat, with the two keepers in the stern, while the other two crews took their places in the water, one on each side of the boat, watching for an opening in the debris-filled surf.

Finally, an opportunity presented itself, and the lifeboat was shot out, the oarsmen bending their backs with a will. The spectators on the beach, even the surfmen themselves, hardly believed that the little surfboat would be able to get through. But by dint of powerful oar work and skillful handling, the craft was brought around to the bar and sufficiently close to the wreck for the lifesavers to make out the crouching survivors.

All around the *Abbott*, however, a jumble of spars and timbers made it impossible for the surfboat to get through to the ship's side. The surfmen tried to force their boat through this network of wreckage but finally were obliged to return to the beach, their strength spent, their lifeboat badly battered.

Scarcely had they landed when the cabin of the *Abbott* toppled into the sea and drifted slowly toward the beach. The surfboat was manned by a fresh crew. This time, the lifesavers managed to get in close to the men—one of whom floated on a hatch covering, the other four still atop the cabin.

All five were pulled into the lifeboat and quickly brought to shore. One of the men, Frank Laven, had a fractured skull and died before medical aid could reach him. The rest were in fair shape, considering the terrible experience they had undergone.

Captain Hawkins said in his letter to headquarters: "With the mass of wreckage in the water being tossed around in all directions, I do not see how the lifesavers launched their boat at all, but they did, and even then, they could not get to us. Finally, when the cabin top broke adrift, they launched their boat again when no man could have expected it. I did not think it possible for them to get to us, but somehow they did, and got us ashore, and I think it a miracle that I am alive to tell the tale. No men could have done more than the lifesavers did."

Chapter III

THE COAST GUARD IN THE GREAT WAR

The rescue of the survivors from the storm-beaten *Reppard*, *Priscilla*, and *Abbott* was another skirmish in the eternal war the Coast Guard is waging against the sea, a skirmish in which beach carts, lines, and the strength of human endeavor were the chief weapons.

But sometimes, there comes an occasion for the Coast Guard to lay aside these weapons and pick up others of a deadlier nature—the weapons of war.

The Coast Guard was created by the act of January 28, 1915, combining the then-existing Revenue Cutter Service and the Life-saving Service, and it was expressly stipulated in that act that the Coast Guard should constitute a part of the armed forces of the United States, to operate in peacetime under the Treasury Department, and under the Navy Department in time of war, or when the president shall so direct.

The Revenue Cutter Service was founded in 1790, some years before the establishment of the United States Navy. Since that time, in cooperation with the Navy, it has served with marked distinction in every war in which the United States has been engaged. History is replete with incidents of service rendered in defense of the country.

In the sea-going branch of the Coast Guard, the cruising cutters maintain a high degree of efficiency in drills with rapid-fire guns, and thorough naval discipline is maintained at all times. In the Coast Guard stations along the coast are warrant officers and enlisted men long trained in maintaining a careful watch over the seacoast and the movements of vessels. Any unusual occurrences within range of observation are at once transmitted over the Coast Guard communications system to Washington.

All the commissioned officers in the Coast Guard, with the exception of a few staff officers and others who have come up from the ranks through sheer force of ability, are graduates of the United States Coast Guard Academy, an institution devoted exclusively to the training of future officers for the Coast Guard, and in which a course of instruction, very similar to that pursued at the Naval Academy, is given. These men were particularly adapted for the work that was to fall to them during the Great War, namely, the handling of vessels designated for escort duty.

In the spring of 1917, when impending trouble with Germany became increasingly apparent, a conference between officers of the Navy and the Coast Guard was held, and a definite plan evolved for the merging of the two related services. Upon notification by dispatch of the declaration of war, each unit was to report to the appropriate superior officer, and provision was made for the immediate absorption by the Navy Department of Coast Guard cutters, stations, and personnel. This plan was known as Plan One.

On the morning of April 6, 1917, a dispatch was sent to each unit of the Coast Guard, stating simply, "Plan One. Acknowledge."

It was war!

Immediately, the Navy was augmented by the addition of 223 commissioned officers, 6,000-odd enlisted men, and fifteen cruising cutters, besides smaller vessels. It is an impressive fact that a large proportion of Coast Guard officers was assigned to command posts. Others not assigned command served with distinction aboard transports, cruisers, cutters, and patrol vessels. Six officers performed aviation duty, two of them being in command of air stations, one of which was in France. It is interesting to note here that one of the pilots of the famous NC-4, the hydroplane that made the first trans-Atlantic flight, was Lieutenant Commander E. F. Stone, a Coast Guard officer.

In August and September 1917, six Coast Guard cutters, the *Ossipee*, *Seneca*, *Yamacraw*, *Algonquin*, *Manning*, and *Tampa*, sailed for European waters to join the naval forces there. These vessels constituted Squadron 2

COAST GUARD PLANE DROPPING HURRICANE WARNINGS

COAST GUARD PLANE RESCUING CREW FROM BURNING BOAT
Removal of crew at sea.

COAST GUARD CUTTER *TAHOE* PASSING ICEBERGS
Cutter *Tahoe*, on international ice patrol, photographed next to two small, flat topped bergs.

ICEBERG FROM THE CUTTER *TAHOE*

of Division 6 of the patrol forces of the Atlantic Fleet. Their base was on the famous Gibraltar, and throughout the war they escorted hundreds of vessels between Gibraltar and the British Isles and performed other escort duties in the Mediterranean. Other cruising cutters performed important duties in home waters, in the Caribbean Sea, around the Azores Islands, and off the coast of Nova Scotia.

The cutter *Tampa*, under the command of Captain Charles Satterlee, headed for the port of Milford Haven, Wales, on the evening of September 26, 1918, having completed her mission as an escort for a convoy from Gibraltar to the British Isles. The night was black, and a slight rain was falling, being driven before a heavy wind. At 8:45 P.M., a loud explosion was heard by people aboard other ships of the convoy, and later, when the *Tampa* failed to arrive at her port of destination, a search by naval destroyers and British patrol vessels was instigated. A few bits of floating wreckage were found, which were identified as being from the *Tampa*. Two bodies, clad in naval uniforms, both unidentified, were picked up.

There is no definite knowledge as to what happened to the *Tampa*, but it is generally believed that she was sunk by a German submarine. Later, the U-53, a German submarine, claimed to have torpedoed a United States ship of the *Tampa*'s description, and listening stations along the coast had detected the presence of an enemy submarine in the vicinity of the place where the explosion was heard. There was not one survivor from this disaster—115 men perished that stormy night, 111 of whom were Coast Guardsmen.

This was, with the exception of the loss of the *Cyclops*, whose fate has never been determined, the greatest loss of life suffered by any American naval unit. Admiral Sims, the American ranking naval officer in European waters, received numerous letters and messages of sympathy from high officials of the governments of the Allies, chief among which was the letter from the British Admiralty:

"Their Lordships desire to express their deep regret at the loss of the USS *Tampa*. Her record since she has been employed in European waters as an ocean escort to convoys has been remarkable. She has acted in the capacity of ocean escort to no less than eighteen convoys from Gibraltar, comprising 350 vessels, with the loss of only two ships through enemy action. The commanders of the convoys have recognized the ability with which the *Tampa* carried out the duties of ocean escort. Appreciation of the good work done by the USS *Tampa* may be some consolation to those bereft, and Their Lordships would be glad if this could be conveyed to those concerned."

Today, when the name of the old *Tampa* is mentioned, Coast Guardsmen lift their heads a trifle higher, and unconsciously their shoulders are squared. It is a matter of solemn pride to the personnel of the Coast Guard to know that the service rendered by the men of the *Tampa* in dying for their country will always be remembered in the naval annals of this country.

On April 16, 1918, the Coast Guard cutter *Seneca* left Milford Haven, Wales, with a convoy of merchant vessels for Gibraltar. In the evening of April 23, the danger zone patrol detachment from Gibraltar came out to meet the convoy. This detachment included the British patrol sloop, *Cowslip*. On the morning of April 28, at about 2:45 A.M., the *Seneca* heard a loud explosion and the *Cowslip* began at once to make distress signals. Captain Wheeler, in command of the *Seneca*, immediately headed his vessel for the *Cowslip*, despite the fact that the torpedoed vessel signaled frantically, "Stay away. Submarine in sight port quarter." The Coast Guard craft circled the *Cowslip* to a position on her port quarter and, joined by the destroyer *Dale*, took up a search for the submarine.

In wartime, it is an established doctrine that when a vessel has been torpedoed, other vessels in the vicinity are not to risk their own destruction by hanging around the scene to pick up survivors. With all justification, therefore, Captain Wheeler might have headed away rather than risk his own vessel. But the Coast Guard has a doctrine, also—to save lives whenever and wherever possible. Captain Wheeler meant to do just that.

Three times the Coast Guard officer stopped his vessel out on the rolling water, with an enemy submarine lurking about, and sent small boats across to the sinking ship. The difficulty the men in those small boats encountered is a sea saga in itself. Coming up alongside the *Cowslip* in the pitch darkness required strong hands and cold nerves. But these men were Coast Guardsmen, trained and particularly adept at this sort of seamanship.

The *Seneca* rescued all the survivors aboard the *Cowslip*—the commanding officer, another commissioned officer, and seventy-nine enlisted men. All the wardroom officers, one enlisted man, and a mess attendant had been killed when the torpedo struck squarely in the wardroom while the officers were at mess.

For his courageous work, Captain Wheeler received a commendation from Admiral Sims and from the British Admiral at Gibraltar. The little *Seneca* became the toast of the fleet.

On September 16, 1918, the *Seneca* was acting as ocean escort for a convoy bound from the British Isles to Gibraltar when the *Wellington*, a British collier and a unit of the convoy, was torpedoed by a German submarine. The explosion tore away the vessel's forefoot, and with her Number 1 hold flooded, she was immediately abandoned. From one of her lifeboats, the collier's captain signaled the *Seneca* that the *Wellington* would probably float but that his crew refused to go back aboard her, as they believed that the submarine would, in all probability, make a second attack.

On the *Seneca*'s bridge, Lieutenant F. W. Brown, U.S. Coast Guard, the cutter's navigating officer, read the message and immediately asked and obtained permission from the commanding officer to take a volunteer crew aboard the *Wellington* and attempt to work her into port.

Practically en masse, the *Seneca*'s crew volunteered for the dangerous duty. Brown went among them, picking his men, until finally, he had a full complement—Machinist William L. Boyce and eighteen enlisted men, most of whom were petty officers. As the crew from the *Wellington* came alongside and scrambled aboard the *Seneca*, the volunteers from the cutter dropped into their boat and shoved off for the *Wellington*.

Once aboard the collier, lookouts and gun crews were assigned and ammunition was broken out. It seemed highly probable that the submarine would return for a second attack, and it was Brown's intention to be well prepared.

Shortly after the Coast Guardsmen came aboard the *Wellington*, a second boat, bearing the *Wellington*'s master, the first and second mates, and eleven of the crew came alongside and climbed aboard.

The master told Lieutenant Brown that he could not stand by and see others do his own duty. Brown at once offered the captain the command post, but the Englishman refused, saying that he was content to serve in a lower capacity. He was at once assigned as the ship's first officer.

Lieutenant Brown and his gallant little band were now on their own. The *Seneca* had been compelled to leave to protect the remainder of the convoy, but she had sent out a call for assistance in code for the *Wellington*.

Machinist Boyce at once made an inspection of the *Wellington*'s engine room and, shortly afterward, reported to Lieutenant Brown that there was steam up and that with only a slight repair job on the air pump, the vessel would be ready for sailing within the hour.

At 12:50 P.M., the *Wellington* started ahead at a slow speed for Brest. The Coast Guardsmen settled down to the precarious task of bringing a torpedoed ship and its valuable cargo through a sub-infested sea. Lieutenant Brown assigned all men who could be spared from necessary duties to making large life rafts, and these were placed out on deck so that they would readily float. A man stationed in the Number 2 hold with instructions to take regular soundings reported three and a half feet of water but added that it did not seem to be rising. As long as the sea was moderate, the pumps would, in all probability, keep ahead of the leak. If the sea kicked up—well, that was the chance Lieutenant Brown and his men must take.

The men were on constant duty. In the engine room the black gang had to be relieved by seamen who had never passed coal before. There was no murmuring, no protest. Their engine room watch over, the men returned to the deck and took their turn at gun watch.

In his official report, Lieutenant Brown comments at length on the behavior of his crew and the diligence they exhibited in the most trying of circumstances. He concludes it with a line: "I consider it an honor to have served with such men."

All went well throughout the afternoon, but at sundown, the wind rose quite suddenly and the sea began to make up. The *Wellington* was loggy and heavy; considerable difficulty was experienced in keeping her on course. At last it became necessary to stop the vessel, as she had persisted in bringing her head into the sea, thus shipping heavy water through her gaping bow.

Unceasingly alert and resourceful, Lieutenant Brown attempted to swing the stern of the ship around so that he might keep way on her by going stern first. The rapidly increasing wind and sea defeated his purpose, and he at once realized the necessity for rigging a sea anchor. The collier, however, was a steel ship, lacking in available material for the construction of a drag that would have pulled the ship's bow around. Brown now thought of cutting the anchor cables, relieving the ship of her anchors and thereby lightening her. But by this time, she had lost all steerage way; heavy seas were sweeping across her decks, making it impossible for anyone to get forward to the chain locker.

It seemed obvious to everyone aboard that little chance remained of getting the collier into port. Indeed, it appeared now a matter of saving themselves, which in itself loomed as an almost hopeless task.

There was only one lifeboat, ready for lowering, but the acute list and the heavy rolling of the collier threatened to throw this craft under the sea. Nevertheless, it offered their only chance of abandoning the *Wellington*. Lieutenant Brown massed all hands abreast of the lifeboat, all except two men at the pumps and the radioman, M. S. Mason, who was at his controls trying to contact the *Seneca* or some other vessel. Seven men of the collier's original crew, together with a Coast Guardsman especially assigned to unhook the boat falls, were ordered to get into the lifeboat. The *Seneca's* men stood by to lower away.

The *Wellington's* first officer, in charge of the lifeboat, had orders from Lieutenant Brown to work his way back under the rail in order to allow the others to abandon ship. Unfortunately, however, one of the Englishmen, fearing that the lifeboat would be crushed against the ship's side by the seas, cut the line and the tiny craft drifted away into the night.

Efforts of the first officer to get back to the *Wellington* met dismal failure, and the only Coast Guardsman in the boat finally stood up and yelled: "Lieutenant! We can't get back! These men don't know how to row!"

Aboard the *Wellington*, a few of her own men and the *Seneca's* volunteers were thus left with nothing to rely on but life rafts that they had constructed earlier. Lieutenant Brown decided to stick to the sinking vessel until the last moment; he knew that the Navy destroyer *Warrington* was proceeding at full speed for his position, which had been previously radioed. He ordered Mason, the radioman, to notify the destroyer that the collier's only lifeboat was adrift and to request them to pick it up. This was subsequently accomplished, but not without dangerous incident: Not only was the lifeboat crushed in the operation, but a boat from the destroyer was broken in two by heavy seas while launching.

Back on the *Wellington*, all planking and material that would float were collected along the side of the ship so that the men might have something to cling to once she was abandoned. The *Warrington's* lights were in sight now, and Lieutenant Brown signaled with a small flashlight that he was compelled to abandon ship immediately and asked the destroyer to work in as close as possible in order to pick up the crew.

Meanwhile, the *Seneca's* volunteer crew had lowered several life rafts to the water and had made them fast with lines so that there would be no difficulty in finding them in the dark. Brown gave the order to abandon ship and stood by until the last man had plunged over the side. Then he signaled with his light that his men were in the water. The collier was now settling by the head and turning over slowly. As Brown sent his last message, he was forced to climb over the rail and stand on the inclined side of the doomed vessel. As he stood there, flashing his last appeal, the boilers of the *Wellington* apparently exploded, and the ship seemed to rise from the water and then settled down. Thrown into the sea, Brown swam with the strength of desperation to avoid the tremendous suction.

This was around 4 A.M., in velvet darkness, with a raging gale whipping the sea to fury. Floating wreckage was Brown's salvation—planking on which was draped the figure of a man. At a glance, he realized that such a frail raft would never float with his added weight. Telling the man to compress his lips to keep out the seawater, Brown struck out again.

He saw two calcium lights burning a short distance away and, thinking that they marked lift rafts, swam toward them. Reaching them, he found nothing but the lights and their metal containers; he extinguished the lights so that none of his men might be misled by them. Apparently, his thoughts, even in these desperate moments, were for the welfare of the eighteen men under his command.

At long last he found himself alongside the *Warrington*. A line whipped down to him. He reached out, caught it, and lashed it about his person. Later he declared that he had no recollection of that moment, having been on the verge of losing consciousness.

Commanded by Lieutenant Commander Van der Veer, U.S. Navy, the *Warrington* was maneuvered with rare ability and persistent courage in and out of the wreckage, searching for the survivors. Lighted life rafts and all the available life rings, well lighted, were floated down upon the still-visible and slowly sinking *Wellington*. Even as the men on the destroyer watched, the collier turned keel up and, after a series of explosions, slid gently beneath the waves.

Dawn was breaking now, a muddy, storm-swept dawn, and the wind was bitterly cold. With the first rays of light, men could be seen in the water, clinging to life rafts and pieces of wreckage. The *Warrington* tried desperately to launch another boat, but, as in the first instance, the lifeboat was crushed, and the attempt had to be abandoned. By means of bowlines thrown to the survivors in the water, eight men were drawn aboard the destroyer, one of whom died soon after being rescued.

It was at this time that Seaman James O. Osborne of the Coast Guard, one of the survivors, earned a life-saving medal and a commendation from the Admiral. Osborne swam with a shipmate, Coxswain Peterson, who had been injured and was unconscious, to a life raft, and placing him upon it, held the injured man as well as he could between his feet. The heavy seas washed Osborne and Peterson away several times, but each time, Osborne rescued the injured Coast Guardsman and replaced him on the raft. Finally, Osborne signaled to the *Warrington*: "I am all right, but he's gone unless you come at once." Both men were rescued.

The survivors from the *Wellington* numbered fifteen men, nine of whom were Coast Guardsmen. Eleven men were lost, one of whom was Machinist Boyce, besides five men of the *Wellington's* original crew. Lieutenant Brown, as previously stated, was rescued but almost succumbed to a severe case of pneumonia brought on by his hours of exposure in the water.

Although the Coast Guardsmen did not succeed in what they had so bravely set out to do, they upheld to the fullest extent the time-honored traditions of the Navy and Coast Guard.

In fact, the British Admiralty were profuse in their admiration. They said: "Seldom, in the annals of the sea, has there been exhibited such self-abnegation, such cool courage and such unfailing diligence in the face of almost unsurmountable difficulties . . . . America is to be congratulated."

While these incidents at sea were carving out an enviable record for the Coast Guard, Coast Guardsmen ashore were not idle. Strenuous patrols along the beach were carried out and a vigilant watch for enemy vessels and submarines maintained.

On the night of October 4, 1918, a tremendous explosion occurred in the loading plant of the T. A. Gillespie Company in Morgan, New Jersey. Tons of TNT had been stored at this plant, ready for loading in ships bound for France. At the first explosion, a company of Coast Guardsmen from Perth Amboy, New Jersey, started toward the scene of the disaster. They were joined by other Coast Guardsmen under the command of the captain of the port of New York and immediately began the extremely hazardous work of removing the dead and injured from the scene of the disaster.

From time to time, as new explosions lit the sky, these men were subjected to veritable barrages of flying steel. The air was literally filled with pieces of shrapnel, and the earth shook. It was as if the war in France had spread to New Jersey's peaceful shores.

Coast Guardsmen died that night in the performance of their duty. One man, while receiving instructions from the warrant officer in charge of the rescue party, was beheaded by a piece of shell; another was killed while dragging an injured person to safety.

Apprentice Seaman C. F. Bennett hadn't been in the service long, but he knew its traditions. He was young, almost a boy, and very frightened. Somehow, he found an automobile and drove it time after time through that barrage of steel, narrowly escaping death on each trip, carrying out the dead and wounded and transferring sentries. A door was blown from the machine, and the top and body were riddled with shell fragments, but miraculously, Bennett remained unharmed and carried on until there was no more to be done.

Another party of Coast Guardsmen, under the direction of Lieutenant J. E. Stika, was told that a trainload of TNT lay in the center of the

Gillespie grounds. By some freak of fate, the TNT had not exploded, but Stika knew that it was only a question of minutes before a direct hit would be scored on these cars, the result of which would be untold damage and death wreaked upon the countryside. It was the Coast Guard's job to move those cars.

When Lieutenant Stika called for volunteers, the entire body stepped forward. Only two of these men—Apprentice Seaman J. Grymes and Bugler N. F. Vaceston—had had previous experience with a locomotive. Grymes agreed to serve as engineer and Vaceston as fireman. But a new difficulty arose. The track had been twisted and torn from the roadbed. Before the nine cars of TNT could be moved to safety, those rails had to be straightened and reset. Though the task seemed hopeless, Lieutenant Stika and his men worked feverishly. Fire was creeping along the track, eating its way toward the boxcars.

Somehow, those men accomplished the impossible. The rails were straightened and the track laid. Grymes started the engine and moved slowly along the makeshift track. It would have been a dangerous job with an ordinary load in the boxcars behind, but with TNT as the load, the slightest miscarriage meant death. The cars swayed across the Gillespie grounds and finally out of the yard and into the safety of the hills. Another job well done by the Coast Guard.

On the forenoon of July 21, 1918, the war was brought to the front door of the Coast Guard itself. The tug *Perth Amboy*, with four barges in tow, was moving southward from East Orleans, Massachusetts, when a German submarine popped up on the port bow and opened fire. The tug hoisted distress signals, and a motor lifeboat from the Coast Guard life-saving station Number 40 put out immediately to aid the shelled vessel.

These Coast Guardsmen knew they were risking death. They were unarmed, and it was reasonably certain that the submarine might consider the surfboat fair game for her shells, but the traditions and standards of their service stood foremost—there was a ship flying distress signals; it was their job to go to that ship's assistance.

The surfboat, in command of Keeper Robert F. Pierce, was within a hundred yards of the tug when the shelling ceased and the submarine

sounded. The Coast Guardsmen met the crew of the tug, who had taken
to small boats, and administered first aid to several of the wounded.

This exploit—unarmed Coast Guardsmen defying enemy guns—ex-
cited praise from every section of the country. Newspapers published
detailed accounts. Nor was the publicity without a touch of humor—
cartoons showed the Coast Guard surfboat proceeding to sea, with boat
hooks as weapons, while the submarine was depicted diving to escape the
fury of the Yankee sailors. While this was not quite in accordance with
the facts, the case did, however, awaken a public realization of the Coast
Guard's rigid observance of its primary duty to the nation.

On the afternoon of August 16, 1918, a surfman in the lookout
tower of the Coast Guard station at Chicamacomio, North Carolina,
was watching a steamer beating its way up the coast in the face of a stiff
northeaster. With the aid of the powerful telescope, he could make out
the name on the bow. She was the *Mirlo*, an English tanker.

As he watched, the *Mirlo* suddenly started to zigzag, and within a few
seconds, there came a terrific explosion and the stern of the tanker was
enveloped in black smoke. The surfman knew instantly what had happened. Torpedoed! He sounded the alarm at once, and the Coast Guard
crew, under the command of Keeper John A. Midgett, prepared to get
underway in a power surfboat to the *Mirlo*'s assistance. The steamer was
approximately four miles out, drifting now, with fire shooting up from
her stern.

When the surfboat was within a mile of the burning vessel, they met
one of the *Mirlo*'s boats with six men in it, one of them proving to be
the torpedoed vessel's commander. The officer informed Keeper Midgett
that two other lifeboats had been launched from the *Mirlo*, and that one
of these had capsized in the vicinity of the fire. The *Mirlo*'s captain said
that he was sure that all the occupants of this lifeboat had perished. The
gasoline from the tanker was running out over the sea, and as this became
ignited, a veritable wall of flame covered the water.

Keeper Midgett directed the officer to make for the beach but to
lay to just outside the breakers and await the surfboat's return from the
Mirlo. They were under no circumstances to attempt to go through the

surf, for Keeper Midgett knew that it would take well-trained surfmen to bring that lifeboat through safely.

The Coast Guard boat then headed for the tanker, which by this time was a mass of flame and low in the water. The sea for a hundred yards around the vessel was covered with burning gasoline, with two great towers of flame about a hundred yards apart.

Dense smoke lay low over the scene, and the surfboat had to proceed cautiously while the men wrapped cloth around their faces. Entering the flames, one man was assigned to stand on the bow with an oar and strike the water ahead of the boat to open up a path through the swirling flame and floating, burning debris. Then, the smoke lifted a trifle, and a lifeboat, overturned, was seen between the two pillars of fire, six men hanging desperately to the sides.

The heat was almost unbearable to the Coast Guardsmen, and the smoke, despite the cloth wrapped around their faces, choked them. With eyes red and burning, Keeper Midgett guided the boat through the flames, until they were alongside the overturned craft. The men in the water were quickly lifted up into the surfboat and the journey back was begun. By this time, the rescuers themselves were in bad shape; the man in the bow actually collapsed and had to be replaced.

By careful work, Midgett brought the surfboat out into the open sea and there rested for a few moments while his men regained their strength and recovered from the choking effects of the smoke. The survivors who related their story to the Coast Guardsmen told how the lifeboat had capsized when an explosion from the ill-fated steamer sent an avalanche of flame and sea down upon them, sending ten men of the original sixteen to a blazing death. The others, hanging to the sides of the lifeboat, had managed to escape the flame-incrusted waves by ducking beneath the water. Holding little hope of being saved, these men had debated among themselves the advisability of drowning themselves rather than risking the tortures of being burned alive. They also said that another lifeboat filled with twenty men was somewhere in the vicinity, and it was believed that they had escaped the burning oil.

Cruising around the edge of the inferno, Keeper Midgett at last found this second lifeboat (she was without oars) drifting rapidly toward the pounding surf. The Coast Guard craft took this lifeboat in tow and headed for the rendezvous Midgett had set with the *Mirlo*'s captain.

When they were still two miles from the beach, the fresh northeaster developed into a gale and the sea rose, making the difficult job of tackling the surf doubly dangerous. Keeper Midgett knew that they would be obliged to go in through the surf while there was still light, but it was already growing dark.

The Coast Guardsmen found the *Mirlo*'s lifeboat, with the steamer's captain and six men in her, safely anchored on the edge of the breakers. Taking a safe load of these survivors, Midgett started his first trip through the surf. Riding the breakers with superb skill, he made three trips back and forth, landing in all thirty-six men from the *Mirlo*, which had long since gone down.

Shortly after the war, another example of the caliber of Coast Guard officers and enlisted men, and of the courage and high degree of seamanship that may be expected of them under adverse conditions, occurred when the squadron under the command of Captain H. G. Hamlet, USCG, ran into heavy weather off the coast of France.

Early in the morning of April 27, 1919, the *Marietta*, with the USS *Teresa*, the *MacDonough*, and nine other small vessels sailed from Brest, France, for Hampton Roads, Virginia. These nine small boats, with the exception of the *Rambler*—a converted yacht of considerable age—had been, at one time, small wooden fishing boats.

Captain Hamlet, in command of the *Marietta*, was the senior officer present and commanded the entire squadron. The weather, which had been ideal at the moment of sailing, kicked up as the day wore on. Seas began to make up, and the going got rough. It was close to 1 P.M. when the *Rambler* broke out a signal, "Man overboard!"

The convoy decreased speed, dropped rafts and markers, and began circling in an effort to pick up the man. The wind, by this time, had increased to a gale, and heavy seas made the rescue work hazardous for the flimsy vessels. An hour later, one of the minesweepers, the *Courtney*, reported that she was leaking badly.

Captain Hamlet ordered her to turn and head for Brest, with the *Mc-Neil*, another mine sweeper, acting as her convoy. But when the weather

CRUISING CUTTER—BUILT 1937

Displacement, 2,000 tons; length, 327 feet; beam, 41 feet; draft, 12 feet and 6 inches; power, geared turbine; maximum speed, 20 knots; maximum cruising radius, 12,300 miles; guns, two 5-inch, two six-pounders, two 50-caliber anti-aircraft, one one-pounder; personnel, 16 officers, 111 men.

SAME TYPE OF BOAT LOOKING AFT

Tampa towing disabled vessel.

SERIOUSLY BURNED MAN TRANSFERRED FROM LIFEBOAT TO
COAST GUARD PLANE

Transfer at sea from lifeboat of steamship *Samuel Q. Brown*, to Coast Guard plane, for speedy
transportation to shore hospital for treatment.

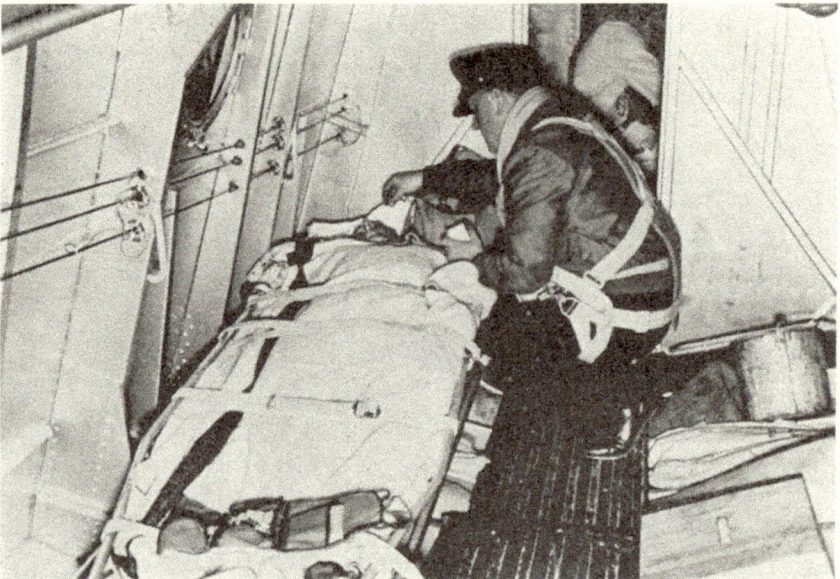

INJURED MAN ABOARD COAST GUARD PLANE

Injured man being transported from ship at sea to shore hospital. The mechanic administering
first aid while radio man in rear compartment sends call for ambulance and doctor to be
waiting. Coast Guard seaplanes are especially equipped for such rescue work.

continued to become worse, all of the converted fish boats were ordered to make for Brest, the *MacDonough* and the *Rambler* to convoy them, leaving the *Marietta* and the *Teresa* to carry on the search for the man lost overboard.

Shortly afterward, word came from the *McNeil* that she was in trouble, and one of the fish boats, then on the way to Brest, was ordered to assist her. Trouble followed trouble. Word flashed from the *Courtney* that she was sinking and needed immediate help.

The *Teresa* and the *Marietta* were forced to abandon the search for the man overboard and proceeded to the *Courtney's* assistance. The *Teresa* was directed by Captain Hamlet to get to the windward of the *Courtney* and make an oil slick, and then drift down upon the stricken vessel and remove the crew. These orders were carried out to a successful conclusion. The *Courtney*, in a sinking condition, was taken in tow by the *Teresa*, and the battered squadron fought the storm toward Brest.

About 9 P.M., a huge wave struck the *Douglas*, wrecking her bridge and injuring several men at the same time. Her seams had opened and water was pouring into the hold. She immediately informed the flagship, *Marietta*, of her condition and asked for assistance.

It was apparent by this time that the *Courtney* could not be saved. The *Teresa* cut her free and proceeded to the *Douglas's* position. By dint of careful maneuvering and skillful seamanship, the *Teresa's* captain placed his ship's bow against the *Douglas's* port side and held her there in that precarious position while the sailors leaped from the sinking vessel's deck to the *Teresa*. In this remarkable maneuver, not a single man was lost, and only one was slightly injured.

In the meantime, the *Courtney* had sunk and the *Teresa* had managed to take the *Douglas* in tow. The little squadron was now in dire circumstances, with one ship already lost and another about to go. So far, however, no lives had been lost, except in the case of the man who had been washed overboard from the *Rambler's* bridge. The storm seemed to be increasing and tremendous seas were running. The *Douglas* twisted and tugged at her lines behind the *Teresa*, and finally, the hawsers parted and the sinking vessel was swept away into the night never to be seen again.

Word flashed from the *James* that she was also in trouble—a flooded engine room and a stove-in bow. The *Marietta* headed for the *James's*

position and, arriving there about 11 P.M., found a tug and two destroyers which had been sent out from Brest to assist the storm-whipped convoy. While the two destroyers and the *Marietta* stood by, laying oil on the
water to smooth the huge waves, the tug managed to get a hawser across
to the *James* and, as daylight broke over the troubled sea, started to tow
her toward Brest.

To add to the situation, Captain Hamlet's flagship, the *Marietta*, ran
into trouble when one of her tail shafts was broken. At the same time, a
plate had become sprung, and water had poured into the engine room,
putting out the fires under one boiler and placing the little gunboat in a
precarious condition.

To top it all off, the *James* parted her lines and broke away from the
tug that had had her in tow, and was rolling in the trough of the sea, her
superstructure a mass of wreckage.

Her commanding officer informed Captain Hamlet tersely: "Do not
believe the *James* will hold together much longer. Men are exhausted
from work at pumps. Would suggest removing crew at once."

Captain Hamlet laid his plans quickly. The destroyers were standing
by, ready to offer any assistance of which they were capable. But Captain
Hamlet knew from past experience that the deck of a destroyer is no
place from which to effect a rescue in heavy seas. A destroyer is too low
in the water; it is practically impossible for anyone to remain on the deck
while these vessels are underway in rough weather.

The Coast Guardsman ordered the destroyers to move around and
form a lee for the crippled *James*, and drift oil down upon her. When this
was done, Hamlet brought his ship around and moved toward the *James*.
It was difficult maneuvering with one engine. A slight error in judgment,
hesitation for just a fraction of a moment, and the *Marietta* would be in
the trough of the sea, an easy prey for mauling waves.

But the Coast Guardsman made no errors, nor did he hesitate. He
brought the *Marietta* broadship to the *James*, and just at the right moment, the order was given, and a Lyle line-throwing gun barked. The line
went sailing into the rigging of the crippled ship. It was drawn in to the
vessel in a few minutes and fastened to a life raft, which was lowered over
the side.

Three men lashed themselves to the life raft and were drawn through the churning water to the side of the *Marietta*, where they were hoisted aboard by means of bowlines. Time after time this operation was repeated, while the *James* settled lower into the water. Finally, three hours later, the last man of the crew of forty-seven was drawn safely aboard the *Marietta*. Within a few minutes, the *James* turned over and sank.

When it is remembered that not a man, except the sailor lost from the *Rambler*'s bridge, died in the sinking of the three ships, despite the fact that a gale of almost hurricane force was blowing, one must realize that here was seamanship seldom equaled and certainly never surpassed. Captain Hamlet supervised the operations from the bridge of the *Marietta*, and every life saved in those trying hours can be credited either directly or indirectly to his ability as a seaman. Perhaps it is not amiss here to mention that Captain Hamlet was afterward appointed Admiral Commandant of the Coast Guard, after having served several years as the superintendent of the Coast Guard Academy at New London, Connecticut.

Chapter IV

THE *MORRO CASTLE* DISASTER

Thoughts of war and disaster were far from the minds of the 549 passengers and crew of the Ward Liner *Morro Castle* on the morning of September 8, 1934, as she plowed her way toward New York. Most of the passengers were still asleep; others were up, making ready to disembark as soon as the ship docked. Storm warnings threatened bad weather up and down the coast, and around 2:30 A.M., the sea was making up rapidly. The *Morro Castle*, however, hoped to beat the storm, as she was making twenty knots and was due in New York within two hours.

George W. Rogers, the chief radio operator, was in his cabin preparing to go to bed when someone knocked hurriedly on the door and entered before Rogers could reply.

It was the third assistant radioman; his face was white and drawn, and he had difficulty speaking coherently.

"Chief! You'd better come up to the radio room! The whole ship's afire!"

Leaping from his bunk, Rogers donned some clothes and dashed for the passageway. The ship was in an uproar. Clad in nightclothes, a woman near the lounge door was screaming hysterically. Seeing Rogers, she lunged for him, clutching his arm desperately.

"It's not so, is it? Tell me, please! The ship's not on fire, is it?"

"Madam, I don't know. Perhaps, it's not as bad as it seems. You'd better go to your cabin and get your things. Try to be calm."

She glanced at him for a moment, seemed to gather courage from his face, then nodded and turned abruptly away.

Rogers shook his head and leaped for the stairs. On B deck now, he ran into a wall of smoke, through which he could see the red flames licking and spitting. People bumped into him, screaming, pushing—praying. Men fought to keep their wives and children around them, strove to calm their loved ones with voices that shook and broke.

Suddenly, out of the smoke and flames, a boy and girl loomed, hands clasped. They were scarcely out of their teens, the boy dark and slender, the girl blonde and tiny. Each wore a life preserver.

The boy said calmly: "You're an officer, aren't you? We're going to jump. How far are we from the beach?"

"Five or six miles," Rogers said, "I don't know. Maybe eight. But you should wait for a boat."

"No," said the boy, "we're going together. In a boat we might become separated. We're not afraid, are we, Judith?"

The girl smiled. "No, we're not afraid—together."

They were gone in an instant.

Rogers made his way to the radio room. The smoke was hanging in wreaths throughout the ship. Near the radio room, in the lounge, the flames roared. The radioman knew that nothing could save the ship now; there remained but one thing to do—summon help without delay.

The radio room was also on B deck, and Rogers knew that the fire would soon cut the cables that supplied the main transmitters with power. That would mean relying upon a small emergency transmitter for sending out a distress call.

The second assistant operator, Alagna, was on watch when Rogers came into the radio room. Alagna said: "The nearest ship is the *Luckenbach*, chief. I heard him talking a few moments ago." He got up and handed the phones to Rogers.

Nodding tersely, Rogers put the phones over his ears. Smoke was piling into the radio room, the steel beneath their feet growing hot. The chief operator glanced beyond the doorway, where flames were crackling.

"Get a release on an SOS from the captain," he said to Alagna. "No time to waste! I must contact some ship!"

When Alagna had gone, Rogers heard the *Luckenbach* asking WSC, the Tuckerton, New Jersey, radio station, if he had received a report that a ship was on fire at sea. Evidently, someone had sighted the flames from the *Morro Castle* and had put a report on the air. Having received no such report, WSC informed the *Luckenbach* to that effect.

Rogers decided to send out a standby signal so that all ships and stations within range would be waiting for the distress call once authority was received from the captain to send one. Accordingly, he switched on his transmitter and pounded out the following:

CQ CQ de KGOV QRX QRX

Within a few seconds after this signal went out, the air became still; the operators were obeying Rogers' instructions to stop sending and stand by.

At 3:20 A.M., Alagna returned from the bridge. Lips tight, he flashed a look at Rogers.

"Hell, chief! The captain refuses to send out a distress call at this time. Claims there's still time—"

As the second assistant spoke, the power went off, plunging the radio room into darkness. The fire had reached the cables—all power was cut from the main transmitters!

Grabbing a flashlight, Rogers whirled to the emergency transmitter. Installed for just such an emergency, this equipment was worked from a bank of batteries outside the radio room. If the fire had not reached them, there might still be a chance—"Here, Alagna!" Rogers snapped. "Hold this flashlight while I time the transmitters!"

"Make it snappy, chief," Alagna whispered through tight lips. "We can't stay here long! Look at that paint on the bulkhead!"

Under the heat, the paint was peeling away from the steel; smoke and flames billowed through the door.

Feverishly, the radioman adjusted his transmitter, praying that the flames had not yet exploded the batteries. He pressed the key and was rewarded when he saw the antenna current meter flicker upward. He was on the air again.

Once more, he whipped out the standby signal. The heat was unbearable, the smoke stifling. He pushed Alagna to the door.

"Tell the captain I can't keep this equipment on the air much longer. Tell him to release an SOS, or by God, I'll release it myself!"

Alagna made his way through the smoke and fire toward the bridge. The flames had gained such headway that he had considerable difficulty in finding his way. Finally, he located the captain on the bridge and shouted in his ear that Rogers would be burned alive in the radio room if the captain did not immediately release the distress call. His face white and drawn, the captain seemed to ponder, then suddenly, he nodded.

"Release an SOS. Our position's twenty miles south of Scotland Lightvessel."

Fighting his way back to the radio room, Alagna shouted over the roaring flames to Rogers: "All right, chief! Send the SOS! Position's twenty miles south of Scotland light!"

Rogers nodded grimly and turned to his key. He had wrapped a towel around his face so that he could breathe in the stifling smoke. He had propped his feet on the table, the steel deck being too hot to stand on. The curtains on the port side of the radio room were now in flames, lending to the scene a ghastly red glow. The distress message flicked out on the air:

SOS SOS
DE KGOV ON FIRE 20 MILES SOUTH OF SCOTLAND
LIGHT
NEED ASSISTANCE

As Rogers finished this transmission, the small generator went out. Eyes blank and hopeless, the two radiomen stared at each other.

"The batteries—" whispered Alagna.

Rogers shook his head. "No. They'll make plenty noise when they go."

He groped his way over to the panel where the leads from batteries were fastened to the switch. While Alagna held the flashlight, Rogers explored the panel and found that the intense heat had melted the solder in the lugs, allowing the leads to slip out and break the connection. Having bent the lead around the lug to the best of his ability, he staggered back to the table.

Again, the small generator hummed and Rogers flung out his frantic call for help for the second time. But no more. With a terrific sound, the batteries exploded. The last means of communication was out of commission!

"Come on, Rogers!" Alagna husked through parched lips. "You can't do anymore! Let's scram!"

But Rogers did not hear. He was slumped in his chair, unconscious; fumes from the exploding batteries had overpowered him.

Picking him up, Alagna struggled to the door. The steel was blistering the assistant radioman's feet. The heat and fumes were almost overpowering. Somehow, he managed to get out in the passageway. Here, the air was a trifle better, and Rogers revived somewhat. With Alagna supporting his chief, the two radiomen made their way to the bridge. It was deserted. Their job finished, they went down from the bridge to the forecastlehead and were subsequently rescued by a Coast Guard surfboat. They had done the best they could for the hundreds of passengers aboard the *Morro Castle* and had upheld the finest traditions of the sea. . . .

Apparently, the *Morro Castle* caught fire at about two-thirty in the morning, the conflagration starting in the lounge room of B deck. It spread so rapidly that at 4 A.M., the entire ship from the bridge to the stern was a blazing inferno. Passengers had no choice but to throw themselves into the water and pray that rescue ships would arrive soon enough to save them.

On watch in the lookout tower of the Shark River Coast Guard Station at this time, Surfman Stephen M. Wilson sighted a red glow in the sky and, a few seconds later, made out the shape of a vessel in flames. He called the officer in charge of the station, Chief Boatswain's Mate M. M. Hymer, and reported the fire. Hymer immediately mustered his crew and, after reporting to the commander of the Fifth Coast Guard District, proceeded to sea in the station's motor surfboat. He had left orders for Surfman Leonard M. Eno to follow him at daylight in the picket boat CG 831.

Bringing his surfboat within a quarter of a mile of the burning vessel, Hymer and his crew could feel the heat from the flames. "Good Heavens, chief," a surfman said hoarsely, "she's a passenger liner!"

Even as he spoke, a cry sounded from a point off the bow. An object was seen bobbing in the water. Nearer the ship and clearly visible in the glow from the flames, other objects dotted the sea.

The Coast Guardsman headed his boat in closer and ran up alongside the man in the water. The surfman reached down and pulled him aboard. The man, past middle age, was clad only in nightclothes with a preserver fastened about his waist. Looking at his rescuers through glazed eyes, he said, "Thank God you came! Hundreds of people are dying, burning to death!"

The Coast Guardsmen counted nearly two hundred persons in the water, some with life preservers on, others swimming desperately. The surfboat moved in among them, picking them up singly and sometimes in groups. People could be seen leaping from the ship, which was covered by flames from bridge to stern. Hymer loaded the surfboat to the danger point. Seeing the lights of a steamer about a mile away, he turned and headed for her.

She proved to be the *Luckenbach* and was already preparing to lower her boats when the Coast Guard surfboat drew alongside. Hymer transferred the survivors, at the same time requesting *Luckenbach's* captain to move in closer so that no time might be lost in getting boats to the *Morro Castle*. Then wasting no time, the Coast Guardsman headed his craft back to the scene of horror.

At 5 A.M., the picket boat from Shark River Station, Surfman Eno in charge, arrived and at once began to pick up survivors from the water. Handling his boat with superb skill in the rising seas, the surfman rescued thirty people. Stephen M. Wilson, the surfman who had originally sighted the *Morro Castle*, dived time and time again from the picket boat. Swimming to survivors who were too weak to help themselves, he assisted many of them to the picket boat.

The last person saved by the picket boat was a young girl who had been supported in the water by a youth. As the craft was now dangerously overloaded, the Coast Guardsman was obliged to proceed, telling the youth that he would have to wait for the surfboat, which was even then approaching from the direction of the *Luckenbach*. Having a life preserver on, the youth was in no immediate danger.

"Right," he said, smiling. "Just take care of my wife. She's fainted."

As the picket boat moved away toward the steamer *City of Savannah*, which was standing by taking survivors aboard, one of the surfmen bent to examine the girl.

"She's dead," he said quietly.

The surfboat under the command of Boatswain Hymer continued to pick up the *Morro Castle's* survivors and, in addition, towed the *Luckenbach's* lifeboats back and forth from the *Morro Castle*. All in all, the Shark River boat rescued eighty people, besides picking up numerous bodies.

In most instances, the survivors were too shocked and stunned to voice opinions about the disaster. A few told how the fire had suddenly broken out; how, in a few moments, the flames had seemed to cover the entire ship. Others related stories of horror aboard the *Morro Castle*, how parents screamed for their children and how a woman had dived back into the flames for her dog. They spoke in jagged voices about the utter madness that had reigned on deck; how the flames had whipped about, driving people to the rails, and then into the water to escape the searing blasts.

The surfboats from the Sandy Hook Station, together with the pilot boat from New York harbor, had arrived at the *Morro Castle* just after daybreak, by which time the Coast Guard boats had already rescued more than one hundred and twenty persons and were searching in ever-widening circles around the burning vessel for additional survivors.

On duty some thirty-five miles east of the *Morro Castle's* position, the Coast Guard patrol boat *Cahoone* received the standby signal from the distressed vessel at 3:21 A.M. but, due to trouble being experienced with the intermediate frequency receiver, failed to get the distress call at 3:28 A.M. The operators worked feverishly on the receiver, and at 3:50 A.M., the *Cahoone* was back on the air.

Cape May Coast Guard Radio Station contacted the *Cahoone* at 4 A.M. and notified her that the *Morro Castle* was afire off Asbury Park, New Jersey. The *Cahoone* proceeded immediately and, arriving at 7:55 A.M., joined in the rescue work.

The Coast Guard Cutter *Tampa*, at the dock in Staten Island, did not receive the distress message direct because of the low-power transmitter being used aboard the *Morro Castle*. At 4:36 A.M., orders came for the *Tampa* to proceed, which she did, after having recalled the liberty crew, at 5:40 A.M. Making about eighteen knots, she completed the run in a little over two hours, arriving at about 7:57 A.M., a few minutes after the *Cahoone*.

MOTOR LIFEBOAT AFTER SEARCH ON GREAT LAKES

Racine Coast Guard Station motor lifeboat returning from search for lost fishermen in ice fields, Great Lakes, February 1936. This type of boat is unsinkable.

COAST GUARD BOAT LANDING FLOOD REFUGEES

Scene at Paducah, Kentucky, January 30, 1937.

ICE PATROL VESSEL PASSING BETWEEN ICEBERGS

The Coast Guard patrols the North Atlantic steamer lanes from March to June, warning all vessels of icebergs and ice fields.

The *Tampa*, being the senior vessel present, took command and directed the *Cahoone* and all but one surfboat to work in toward shore, for it was hoped that survivors had themselves drifted or swum in that direction.

Meanwhile, the Coast Guard surf stations along the beach were kept busy. A crew from Monmouth Beach rescued sixteen people from the surf, the surfmen forming a life chain out in the water in order to bring the survivors in. Surfmen of the Squan Beach Station rescued four persons and picked up six bodies; Bay Head Station saved five people, but only after Surfmen Grayson J. Mears and Robert A. Gent had fought through the rolling surf to bring them in. These two men nearly lost their lives in the effort and afterward required medical treatment.

Nearly thirty Coast Guard surf stations assisted in the *Morro Castle* rescue work, together rescuing 173 persons and recovering thirty-eight bodies. The service itself was credited with saving a total of 259 survivors and recovering fifty-six bodies.

Captain W. H. Shea, commanding the New York Division of the Coast Guard, stated in a letter to headquarters: "In the opinion of the division commander, Chief Boatswain's Mate (L) M. M. Hymer was the outstanding man in the Coast Guard in the performance of effective duty during the *Morro Castle* disaster. The fire was reported to him at 3:15 A.M., standard time, and he immediately started in action, as seen from his report. Going to sea in conditions that required and obtained great skill and seamanship in small boat handling, he and his crew in the motor surfboat should receive special commendation from headquarters. Also, Surfman Leonard M. Eno and the crew of the CG 831 should be commended for their fine work. Special mention is made here of Surfman Stephen M. Wilson, who swam to various survivors and helped them into small boats. Signed, W. H. Shea, Commander, New York Division."

The forepart of the *Morro Castle* was comparatively untouched by the flames, the fire having confined itself to the middle and afterpart of the ship. Boatswain Morin, in charge of the Sandy Hook surfboat, informed the *Tampa*'s commander that fourteen men were still aboard the burning liner, having taken refuge on the forecastlehead. Boatswain

Morin was directed to go as close as possible to the *Morro Castle* and take these men off.

After establishing communications with the liner's acting commander, Warms, Boatswain Morin reported to the *Tampa* that Warms had requested a line be put aboard the burning ship and a tow attempted to New York harbor, where fire boats could extinguish the conflagration.

The *Tampa*, with the aid of the New York pilot boat, finally managed to get a running line across to the *Morro Castle*. Now began a real struggle. As the latter's engine room had long since been abandoned and there was accordingly no steam on the winches, the men on the liner's forecastlehead had to pull in the hawser by hand. After an hour of struggle with the wet and heavy line, they managed to get it through the anchor eye and make it fast. Next, the anchors were let go and the *Morro Castle* swung round in the sea, the wind coming across her stern, bringing the smoke and heat down on the fourteen men on the forecastlehead. The *Tampa* immediately took up a strain on the hawser and, slowly, the *Morro Castle*'s nose came around. The wind was directly off the bow now, and a heavy pall of smoke clouded the sky behind the burning vessel. For the men on the forecastlehead, this was a welcome relief, the intense heat being thus alleviated.

There remained nothing more for these survivors to do. Realizing that the increasing gale might shift and drive the flames upon them, the *Tampa*'s commander ordered the Sandy Hook surfboat to go alongside the *Morro Castle* and effect the rescue.

In charge of the surfboat, Boatswain Morin knew that here was a task that would try his seamanship to the limit. The *Morro Castle*, free from the weight of her anchors, was pitching heavily, yawing widely behind the Coast Guard cutter. It was absolutely necessary that the Coast Guardsman approach close enough to the burning ship's side to permit rescue of the survivors as they clambered down a Jacob's ladder.

Actually, there seemed very little chance of doing this without staving in the surfboat against the *Morro Castle*'s blistered side and throwing the Coast Guard crew into the sea. Boatswain Morin was fully aware of these risks, but he had his orders. He maneuvered the surfboat closer and closer to the *Morro Castle* while around him mingling smoke and spray dangerously limited visibility. The men on the forecastle watched breath-lessly, for this might be their last chance at salvation.

The surfboat was close in now. But—suddenly—the *Morro Castle*, like a maddened sea monster, lunged directly at the surfboat. Desperately, Morin whipped his wheel over, his eyes pinned on the mass above him. The distance narrowed; collision seemed inevitable. Then, at the last moment, the *Morro Castle* listed suddenly away and, as though disgruntled at its failure to crush the surfboat, rolled restlessly in the troughs.

Morin brought his craft around for a second attempt. This time, he was more successful and, in a few moments, lay under the *Morro Castle's* rail. He held the surfboat there, with only a yard or so separating him from disaster, until the last one of the fourteen survivors was safe on board. It was 9 A.M.—a gray morning with fine rain drifting in sheets before the increasing wind—when Morin headed away from the burning liner.

A storm, previously forecast, was making up in earnest and the seas were tossing angrily when, soon after Morin had rescued the last survivors, the hitherto mentioned *Tampa* got underway and steered for New York with her tow. She made little headway. The *Morro Castle*, without a rudder or any other means of guidance, kept yawing broadside behind the *Tampa*, stretching the hawser so tight that it droned like a reed in the wind. Anxious eyes aboard the Coast Guard cutter watched that line, waiting for the loud "crack!" that would come when it parted. Once that happened, all hopes of saving what was left of the palatial ocean liner would be lost, for there would be no chance whatever of placing another hawser aboard her.

The *Tampa's* commander requested the New York pilot boat to pick up a line that was dangling from the *Morro Castle's* stern and, in this manner, act as a rudder so that the vessel might be held as well as possible on the course.

This was done, and for a time, it seemed that Coast Guard ingenuity and perseverance would bring the *Morro Castle* into the harbor where fire boats could extinguish the fire. Such good fortune was not to be, however. After a day's struggle with the burning hulk, the pilot boat was forced by the now-raging gale to abandon its post. Aboard the Coast

Guard cutter, it was feared that the hawser would part at any moment, and precautions were therefore taken to prevent injury to the men in the event that the broken line should come hurtling back upon the *Tampa*.

Hardly had this been done when the hawser snapped with a report like that of a three-inch gun. The *Tampa* surged ahead, trembling throughout her frame. Destruction threatened her. Writhing through the air, the hawser struck her stern and fouled the churning screw. The cutter trembled. Bells jangled frantically in the engine room as the commander rang STOP on the telegraph.

Her screw now hopelessly fouled, the cutter took the seas broadside and at once began to drive toward the beach, from which sounded the thunder of the surf. Furthermore, a sand bar was known to bear a cable's length ahead.

In response to orders from the bridge, both anchors were let go in an effort to check the *Tampa's* headlong plunge to destruction. The men wondered: Will she hold?

Those were anxious moments. The anchors dragged. Then, suddenly, they bit, took solid hold, and the cutter lay within a few hundred feet of the beach.

The *Tampa* flashed a call to divisional headquarters in New York, and the cutter *Sebago*, which had been undergoing repairs, was ordered to assist her into port. The *Tampa's* work was done. Along the beach, however, weary patrols from the Coast Guard surf stations continued to search for bodies. This work went on for days until the surfmen were ready to drop from fatigue. Not until hope for the recovery of the rest of the bodies had been abandoned did these men get rest. Some of them had been on their feet for seventy-six hours. The concerted efforts of the Coast Guard units resulted in the saving of 259 lives and the recovery of fifty-six bodies.

Meanwhile, the *Morro Castle* had been driven ashore and had brought up near the pier at Asbury Park, New Jersey. The flames were dying now, but tremendous pillars of smoke still belched from the charred hull— a grim, blackened reminder of one of the worst tragedies in maritime history.

Chapter V

SOS!

An earlier broadcast of the trouble aboard the *Morro Castle* would have resulted, undoubtedly, in many more lives being saved, since it would have given the Coast Guard and commercial agencies more time in which to coordinate their rescue efforts.

In the *Uvira* case, the free use of the radio, coupled with the outstanding alertness of the radio operators aboard the Coast Guard cutter rescue vessel *Pontchartrain* was chiefly responsible for the ultimate rescue of all persons aboard the yacht.

Cape Hatteras, on the North Carolina coast, has long been known as a graveyard of ships. It juts out into the ocean like a huge bow, its dangerous shoals lying in wait for the unwary sailor. During stormy periods, these shoals cause vicious cross seas and hidden currents which exert such a strong pull toward shore that many a superstitious old-timer believed the place accursed.

On the morning of February 13, 1933, the cutter *Pontchartrain* was standing down Thimble Shoals Channel, Lynnhaven Roads, Virginia, on her way out to sea for regular patrol off Cape Hatteras. Weather reports had come in, advising that a storm of considerable intensity had centered in that vicinity. Hatches were battened down and provisions made for rough weather, for the Coast Guardsmen knew what Cape Hatteras was like in a storm.

In the radio room, Shimkus, radioman first class, was on watch. A veteran operator, he knew that stormy weather usually meant work for the off-shore patrolling vessels. The air was filled with the clamoring of ships at sea, some of them asking for weather reports, others sending in their positions. Shimkus copied down the positions as they were sent, for he did not know how long it would be before one of those very same ships might need assistance.

At 11 A.M., a ship opened up suddenly with a shrill insistent note. The signals pounded out over the growl of the sparks. It was an NCU!

NCU is the general call for the Coast Guard. While it does not necessarily indicate distress, it does signify that the transmitting station has a message of utmost importance that he wishes to clear to the nearest Coast Guard unit. This call was from a Navy destroyer, which, after getting Shimkus's answer, shifted to his working frequency and sent the message:

NRUP DE NUSJ HR FOLLOWING MESSAGE RECEIVED
FROM YACHT UVIRA IN POSITION LAT 3 0-12 N. LONG
74-13 W. QUOTE WE ARE TAKING WATER FAST MAIN
BOOM CARRIED AWAY ONE MAN LOST OVERBOARD
PASSENGERS ABOARD NEED IMMEDIATE ASSISTANCE
UNQUOTE UVIRA'S SIGNALS EXTREMELY HARD TO
COPY POSSIBLY DUE TO ANTENNA GROUNDING OUT

Shimkus receipted for the message and called the bridge. In a few minutes, the *Pontchartrain*'s commanding officer, Commander Farley, had the message, and the cutter's speed was increased to full. Chief Radioman Harrison was at the radio direction finder on the bridge, attempting to tune the *Uvira*'s signals in so that radio bearings might be taken on her.

WSC, a powerful commercial radio station at Tuckerton, New Jersey, became aware of the situation almost instantly and broadcast a silencing signal so that communications with the *Uvira* might not be impaired by interference. In the silence that followed, Shimkus heard the *Uvira* repeating her message, begging desperately for immediate assistance. The signals were difficult to read, being broken up and ragged. The radioman aboard the cutter knew what was happening. The distressed ship's antenna was down or else grounding out in some way.

Harrison, at the direction finder, also heard the signals and managed to get bearings before the *Uvira*'s transmitter ceased. These bearings were plotted on the chart and were found to be at a considerable variance from the position given in the message received by the Navy destroyer. The chief radioman immediately checked the bearings, but the same results were obtained. The position given by the Navy ship placed the *Uvira* north of Lynnhaven Roads, while the bearings indicated that she was somewhere near Hatteras.

By this time the cutter was well out in the open sea. Dirty green water broke over her bow, swept along the deck. The *Pontchartrain* was being driven hard, and the sturdy cutter shuddered every time a big wave came aboard, but she kept plowing on, her speed undiminished. Commander Farley decided to ride out the bearings obtained with the direction finder rather than rely upon the position given by the destroyer. The odds were in favor of the message having been garbled on account of the difficulty experienced in reading the *Uvira*'s signals.

In a few minutes, the wisdom of Commander Farley's decision was borne out. The SS *Atenas*, a commercial steamer, called the *Pontchartrain* and reported that her position was three hours away from the *Uvira*. Commander Farley checked the relative positions of the *Pontchartrain* and the *Atenas* and found that the cutter was on the right course. The master of the *Atenas* asked if he should proceed to the *Uvira*'s assistance, and Commander Farley, knowing that minutes might mean the difference between life and death for those aboard the yacht, answered affirmatively. He further informed the *Atenas* that the *Pontchartrain* would arrive at the *Uvira*'s corrected position at about 5 P.M. and requested the steamer to stand by until the cutter's arrival.

By this time, the *Uvira*'s signals had died away altogether. The *Pontchartrain* called her several times, as did the *Atenas*, but there was no answer. Her silence might mean that the vessel had at last been broken up by the force of the seas. The cutter's commander called for every ounce of steam in the boilers, and the *Pontchartrain* hurtled on, straining like a greyhound at the leash. Commercial vessels kept reporting their positions to the cutter, asking if they should proceed to the assistance of the *Uvira*. The *Pontchartrain* informed them that the *Atenas* and herself were the closest vessels to the yacht and additional assistance was not needed.

At 3:30 P.M., the *Atenas* reported that she had sighted the *Uvira* in position latitude 35-51 N. longitude 74-02 W. The yacht was bare of sails except for a ragged strip up forward, evidently a distress flag. She had been taking a terrific beating from the seas, and immediate assistance was necessary if the yacht and passengers were to be saved. The *Atenas* reported her own attempt to put a line aboard the *Uvira* and the defeat of the effort by the gale and treacherous seas. Backing off, the steamer used her transmitter to guide the cutter to the spot.

The *Pontchartrain* sighted the *Atenas* and the *Uvira* at 5:50 P.M. It was growing dark and, to add to this difficulty, snow fell thickly. Visibility was done. Lives had to be saved, and the Coast Guardsmen were used to taking outside chances in their rescue work. These chances had a habit of paying high dividends, as records will show.

Commander Farley threw the annunciator handles down to half speed, and the cutter plowed down upon the *Uvira*. The helmsman was tense, his eyes narrow as he jockeyed the *Pontchartrain* closer and closer to the bow of the yacht. The Coast Guardsmen on the stern of the cutter were ready.

When the bow of the *Uvira* appeared amidships, the officer in charge snapped an order, and the man with the line-throwing gun stood up, gun at his shoulder. Commander Farley judged the distance with a practiced eye and, at exactly the right moment, gave his command. The helmsman spun the wheel. Instantly, the cutter went over on her side, heeled into the wind. The stern passed just under the bowsprit of the *Uvira*, so close that a collision appeared inevitable. Then, the gun echoed hollowly against the roar of the sea and wind, and this time, the line went squarely across the yacht. A man darted out, made it fast. The *Pontchartain* sheared off, lay to a hundred feet away.

The Coast Guard had done its part, but the fight was not yet finished. An eight-inch hawser had to be pulled aboard the yacht and made fast for towing. The freezing weather and the rolling of the *Uvira* made this a Herculean task, but the crew of the disabled vessel knew that the hawser was their only hope of salvation. For an hour and a half, they worked in the lulls between the seas, worked with ice hanging to their clothes and with hands numb with cold. Finally, the hawser was pulled in and shackled to the starboard anchor chain.

At 9 P.M., the *Pontchartrain* got under way with her tow. In order that the yacht would not be subjected to any unnecessary punishment, Commander Farley ordered a speed of 46 rpm for the night. This was barely enough to give the cutter steerage way, but the commander was counting on holding his own until morning, waiting for a lull in the storm.

All that night, the *Pontchartrain* held the bow of the yacht into the seas. A searchlight was kept trained on the *Uvira* and the lookouts were alert for any signal. The vessel appeared to be shipping water continually but seemed to be holding together as well as could be expected.

The next morning the seas had moderated considerably, and the speed was increased to 80 rpm. The *Uvira* appeared to have come through the night fairly well, except that her superstructure was a mass of wreckage. By mid-afternoon, the cutter's speed had been increased to 100 rpm and a course for the Chesapeake light vessel. Then, at 8:20 A.M. on February 15, the *Uvira* was brought into Hampton Roads and anchored.

Ensign Grantham of the *Pontchartrain* went across to the *Uvira* to complete the Coast Guard's report. He found that the yacht carried twenty-two in her crew and three passengers, two of the latter being women. Another passenger, Herbert Kuechenmiester of Chicago, had been washed overboard on the morning of February 13.

The yacht had been bound from New York to Miami and was two days out when she ran into the storm. She had weathered the blow admirably until her motors failed. The crew had then tried to rig the sails in the raging wind, but the canvas was whipped to pieces and the boom carried away. Kuechenmiester was washed overboard at this time while helping the crew. From then on, it had appeared just a question of time until the yacht would go to pieces unless her frantic calls were heard and help came.

The *Pontchartrain* towed the *Uvira* 170 miles under the most adverse weather and sea conditions. The seamanship exhibited in that dramatic rescue has gone down in the annals of the Coast Guard as the most daring and resourceful of all rescue operations. To the Coast Guardsmen who took part in the episode, it was just another job well done, for which they felt undeserving of unusual praise.

Soon after the *Uvira* rescue, a news story appeared in an eastern newspaper stating that the Coast Guard had presented a bill to the owners of the *Uvira* for "towing charges." To say that this was untrue would be contributing some measure of respect to a bald misstatement that deserves no tissue of respectability. A storm of protest came from the Coast Guard. While they wanted no credit for their work, they did resent any such slur as that which had been cast upon them. Ensign Grantham, the only Coast Guard officer to visit the *Uvira*, presented an affidavit to headquarters, denying that any such charge "for towing" had been made while he was aboard. It was never established just where the reporter got the idea that the Coast Guard charges for its work. The service that the Coast Guard affords distressed mariners is free; it is a service unparalleled in any other country in the world.

The sea had been cheated in the *Uvira* case. Twenty-five persons had been snatched from its clutches, but five years later, almost in the same spot, the sea was to score heavily, sending the Greek steamer, *Tzenny Chandris*, with eight of her crew, to the bottom. The *Tzenny Chandris* had recently been bought in the United States by a Greek syndicate, and this would have been her first trip across the Atlantic.

On the morning of November 13, 1937, the *Tzenny Chandris* was a day out from Morehead City, North Carolina, bound for Rotterdam, Holland, with a cargo of scrap iron. She had been bucking an easterly gale almost from the moment she had left Morehead City, and the ship was creaking in every joint with the strain of the seas. There was water in the hold and more pouring in between the plates of the tortured ship. The pumps had been going for hours, but the water was gaining steadily.

The *Tzenny Chandris* was becoming loggy and fast losing steerage way. The master knew what that portended. Once the forward speed of the vessel was lost, she would be at the mercy of the pounding seas, which were now roaring down upon her from a height of nearly forty feet.

The *Tzenny Chandris*'s master gave orders to the radio room that a distress call be sent out. There was nothing left to do. The storm showed no signs of abating and the steamer could not long survive such a beating. Already, the engine room floor plates were covered with six inches of water. Once the water reached the fires under the boilers, it would be all over.

In the radio room, the harassed operator had been trying, unsuccessfully, to contact another ship or shore station. He was not an experienced operator, and this, coupled with the fact that he knew but little English, contributed to the urgency of the situation. The radio equipment was old, the transmitter being of a low-power spark type. The receiver was practically useless since no signals from a great distance could be received with it.

The operator knew that the odds were against him. The knowledge that the lives of nearly thirty men depended upon him brought a sob of hopelessness to his lips. Desperately, he turned to the transmitter. He saw that the power was low, but he had to get out. God, he had to get out. . . .

The Coast Guard cutter *Sebago*, on patrol at sea, had heard a weak spark signal several times during the night, broadcasting some sort of a message, but the signals had been too weak to copy. And the radioman on board the cutter was worried. With a keen intuition born of years of experience, he felt that something was wrong. He swung the dial of the receiver back and forth, searching the air for the guttural moan of the spark signals.

Then, during a momentary lull, the signals came again. This time they were a bit stronger and were fairly readable. It was a message addressed to all ships and read:

CQ CQ DE TZENNY CHANDRIS CQ ALL SHIPS PLEASE
TZENNY CHANDRIS STANDBY WE ARE IN VERY
TROUBLE POSITION PLEASE ANSWER

The message was obviously in broken English, but the radioman knew that it had come from a vessel in distress. He snapped on his transmitter and answered the *Tzenny Chandris* in a slow, even swing, asking for a position and more details. There was no reply.

Poyners Hill Naval Direction Finder Station, on the North Carolina coast near Hatteras, had also heard the message. An attempt was made at that time to secure a bearing on the distressed ship but without success. However, a few minutes later, at 4 A.M., the Greek ship sent out an SOS, the first she had transmitted, and closed off abruptly without having sent her position. The direction finder station managed to get a bearing on this transmission, but due to weak signals, the bearing was considered doubtful. It was forwarded, however, to Naval Radio Headquarters in Norfolk for further transmission to Coast Guard units at sea.

The *Tzenny Chandris* continued to fill the air with SOS from 4 P.M. to 4:30 P.M., after which she was not heard again. Never once did she give her position; thus, invaluable time was lost. Her transmissions were brief, doubtlessly due to power failure, and this fact greatly hindered direction finder work. Her usage of the complete name, *Tzenny Chandris*, instead of a regularly assigned call added considerably to the confusion already existing.

Tuckerton Radio, the commercial station that had done such sterling work in controlling communications during the *Uvira* case, came on the air and began inquiring of ships if they had been in communication with the *Tzenny Chandris* any time prior to the distress. Finally, the SS *Seatrain* told Tuckerton that she had worked the *Tzenny Chandris* at 10 A.M. the day before and that the Greek ship's position at that time had been near Hatteras, bound north. This information was intercepted by the *Sebago*, and the cutter headed full speed for the estimated position of the *Tzenny Chandris*.

The *Sebago* arrived off Hatteras at 9 A.M. She at once began to circle, gradually widening her sweep to take in as much space as possible in the search for the *Tzenny Chandris* or possible survivors. It was a rough job, for the seas were kicking up strongly from the force of the easterly gale. At 9:07 A.M., the *Sebago* heard an oil tanker, the SS *Swiftsure*, talking with a sister ship, the SS *Lucas*. The *Swiftsure* told the *Lucas* that she had picked up a lifeboat with six men at a position approximately forty miles northeast of Diamond Shoals Lightship.

This information was received also by the Coast Guard radio station at Cape May and rushed to the Commander of the Division at Norfolk. The air station was directed by the commander to hold planes in readiness in case the weather moderated enough to permit them to be used in a search for the survivors of the *Tzenny Chandris*, which by this time was known to have gone down.

The *Swiftsure* called NCU, the general call for the Coast Guard, and the *Sebago* answered. For some reason or other, the tanker did not hear the cutter's answer but broadcast the message for all ships:

ALL SHIPS VICINITY DIAMOND SHOALS ONE LIFEBOAT
AFLOAT WITH MEN APPROXIMATELY THIRTY TO
FORTY MILES NORTHEAST OF DIAMOND SHOALS DO
NOT KNOW WHAT NAME OF BOAT THEY SPEAK GREEK
ONLY THE BOAT PICKED UP HAD SIX MEN AND THEY
SAID ANOTHER BOAT WAS AFLOAT WITH FOURTEEN
MEN IN IT

This message was followed by a second:

TZENNY CHANDRIS OF CHIOS GREECE SAILED FROM
MOREHEAD CITY THURSDAY NOVEMBER 11 AND SANK
ABOUT FOUR A M NOVEMBER 13 STOP WE PICKED
UP SIX MEN IN LIFEBOAT THIRTY MILES NORTHEAST
DIAMOND SHOALS LIGHTVESSEL STOP SURVIVORS
CLAIM EIGHT MEN IN WATER WITH LIFE BELTS STOP
WE WILL SEARCH FOR SAME IN VICINITY SIGNED
ALLEN MASTER

By this time, the cutter *Mendota* was on her way out to sea from her base at Norfolk to assist in the rescue operations, followed closely by the patrol boat *Dione* and the cutter *Bibb*. The Naval air authorities at Norfolk called the divisional headquarters and offered the services of seven large patrol planes. This offer was accepted by the Coast Guard, but due to highly unfavorable flying conditions, it was not considered advisable at that time to order the planes into the air.

During that afternoon and night, the cutters *Mendota*, *Bibb*, *Sebago*, and the patrol boat *Dione* plowed the waters around Diamond Shoals, searching for survivors. It seemed a hopeless task, but the Coast Guardsmen held to their work grimly. The weather had moderated somewhat, but the seas were as high as ever, and it did not seem possible that men could live in them. Their search was fruitless, but at 6 A.M. the next day, a message was received from the commander of the division, stating that the seven Naval planes had departed from Norfolk and that the Coast Guard patrol plane, V-126, pilot Lieutenant R. L. Burke, was on its way from the air base at Cape May, New Jersey, to join in the rescue work.

The naval plane 14-P-8 advised the cutter *Mendota* at 10 A.M. that she had sighted bodies floating in the water approximately seven miles from the *Mendota's* position. The plane circled and dived in order to guide the cutter to the spot. Arriving there, the *Mendota's* boats picked up the bodies of three men in life jackets, floating upright in the water. It was obvious that one of the men had died from injuries received in the wreckage; the other two had apparently died from exposure. The carcasses of pigs and other animals dotted the water around the wreckage. Foreign freighters are accustomed to taking with them livestock for provisions since most of the ships are not equipped with refrigeration.

Meanwhile, the Coast Guard patrol plane had arrived and reported for duty to the *Sebago*, which was senior ship. The plane was assigned an area to scout, taking over the work that the Naval planes had been doing. The fuel of the Naval craft was beginning to run low, and they requested permission to return to the Naval Operating Base at Norfolk.

Scouting southward, over an area of perhaps twenty miles square, the V-126 at once began to sight survivors clinging to rafts and bits of wreckage. A group of four or five was holding on to an overturned lifeboat, probably the one that the *Swiftsure* had mentioned. Lieutenant Burke flew low over them, dropping smoke bombs to mark the spot for the cutters.

Then Lieutenant Burke became aware of a terrible menace facing the men in the water. Dark, sinister shapes hovered around the carcasses of the animals from the *Tzenny Chandris*, and Burke knew at once what those shapes were. Man-eating sharks!

The sharks were beginning to leave the carcasses of the animals and forming a deadly ring about the weakened men. Some of the survivors had sticks with which they beat weakly against the water to frighten the man-eaters away, but the sharks would dive rapidly and disappear only to show up a few seconds later, closer than ever to the men. It was only a question of minutes before they would launch themselves in attack.

Lieutenant Burke advised the *Mendota* of the situation and requested that all speed be made to the scene, or the plane would attempt to land in the rough sea and pick up the men. This Lieutenant Burke did not wish to do except as a last resort, for he knew well enough that the plane would never rise from that water, as rough as it was.

Burke managed to hold the sharks at bay by zooming and diving the plane over and at them. The *Mendota* steamed up and lowered her boats while crack sharpshooters lined the rail with .30-30 rifles. Fourteen men were picked up, all in an exhausted state. They had been in the water for nearly thirty hours, their boat having capsized just after leaving the *Tzenny Chandris*. Two bodies were found in the vicinity of the wreckage, and one old man died soon after being brought aboard the *Mendota*.

The survivors were taken below and given hot coffee and medicinal whiskey. One man was given treatment for a bad wound on the heel where a shark had bitten him. This man spoke a little English and he told of the tragedy in shaking, broken tones.

Soon after leaving Morehead City, the vessel had started taking water in the hold. The captain had thought that they would be able to make port, but a heavy blow had caught them while going around Hatteras. The storm had grown worse and so had the condition of the *Tzenny Chandris*. Finally, the captain had called them out on deck and told them that the end of the voyage had come and that all attempts to get help had failed. They were going to abandon ship.

One boat had been put over the side without mishap. The second boat got away, but a big sea capsized it, dumping the men into the angry water. That same sea had come aboard the *Tzenny Chandris*, carrying away ventilators, flooding the engine room, and shifting the cargo. The ship, by this time, had a fifteen-degree list and was listing further. The captain ordered his men to jump into the sea in an effort to save themselves and watched them as they went one by one over the rail.

The rest of the horrible experience was hazy for this survivor. His story, at this point, became rambling, incoherent. He did tell of seeing one of his shipmates being pulled from the raft by a shark, and he himself had been attacked at the same time. This was just before the planes came down through the mists to locate him and his shipmates and end his nightmarish adventure.

The *Mendota* found two more men and another body. In all, the cutter picked up three bodies and saved sixteen men, one of whom died aboard ship. The survivors had been driven nearly fifty miles from the position where the ship was supposed to have gone down; they were in the water for thirty hours. The *Sebago* had been a relatively short distance from the position of the *Tzenny Chandris* and, but for the failure of the freighter's radioman to broadcast his position, might have been able to rescue the entire crew. The bearings obtained during the time the *Tzenny Chandris* was on the air were generally inaccurate. This was due, however, to the weak signals and broad minimum afforded by the ship's obsolete transmitter.

Eight lives were lost in this disaster. It is, indeed, a miracle of modern times that any of the *Tzenny Chandris* crew were rescued. You can charge that miracle up to another miracle—aviation! Had it not been for the Naval planes and the Coast Guard V-126, it is probable that not one of the sixteen men saved by the *Mendota* would have been found alive.

Chapter VI

THE SEA IS A KILLER

The work of the Coast Guard cutters in the *Uvira* and the *Tzenny Chandris* cases was an outstanding example of the seamanship these men are called upon to exhibit in the open sea and the thoroughness with which they do their job. It remained for the salvage work upon the grounded ship *Childar* to bring out the resourcefulness and the tenacity that are traditionally a part of the Coast Guard.

On May 4, 1934, the Norwegian motor ship *Childar*, with a cargo of lumber, a crew of twenty-nine, and one passenger, was bound from Longview, Washington, to South Africa. A heavy southerly gale was blowing, and the weather was thick with the haze from an angry sea. There was a worried look on Captain Matthisen's face as he stood on the bridge with the watch officer and stared across the water.

The weather was bad, but Matthisen wasn't thinking particularly about that. He was an old-time sailor, and he had known the sea in all its moods. Besides, the *Childar* was practically a new vessel, built to stand exactly the treatment she was getting now. There was something else on his mind, some vague undertow of thought that kept pulling his muscles taut and bringing a dry feeling to the roof of his mouth. . . .

The *Childar*'s position at this time was just off the entrance of the Columbia River, a bad spot for any ship in rough weather. Sand spits and shoals lay hidden by the water, forming a deadly trap for the ship blown from its course. With this thought in mind, Matthisen walked over to the chart and checked his position, allowing a comfortable margin to seaward of the shoals. The chart showed him to be in safe waters, but his feeling of uneasiness persisted. The captain started to shrug, then stopped, his eyes wide, his face strangely pale. A hollow roar seemed to reach out from the vastness of the sea.

A moment later, before the men on the bridge could move, the *Childar* crashed against the rocks, and her wrenched timbers screeched against the strain. She was thrown far over on her side; giant breakers raced across her decks, sweeping boats and deck load clean away. Picking himself up from the deck where the force of the impact had thrown him, Matthisen ran out on the wing of the bridge. The *Childar* was rolling forward, floating across the shoals into deeper water. She trembled violently as her hull scraped across the rocks.

It was easy to understand now why the *Childar* had run aground. The wind and sea had driven the vessel toward shore, and this, coupled with the fact that the ship had been navigated by dead reckoning in which a mental slip is often disastrous, had put the *Childar* on the beach. Matthisen stopped thinking about this; there was work to be done if they were to escape with their lives.

The first mate had rushed out on deck, followed by members of the crew. One lifeboat had been left swinging in its davits, and it was imperative to secure this boat before another sea carried it away. That lifeboat would probably mean the difference between life and death for the *Childar's* crew, for even now, it was apparent that the ship could not hold together very long.

Then, out of the murk and roar of the surf, a giant comber rolled down upon the stricken ship. Matthisen shouted a warning from the bridge. Too late! The breaker thundered aboard and buried the vessel under a green torrent of water. It swept the mate and two seamen overboard. Another man was crushed to death between the boat and the deck. Three other seamen had suffered injuries, two of them seriously. The lifeboat itself was a mass of twisted wreckage on the *Childar's* littered deck.

By this time, the ship had been driven far up on the rocks. One end protruded over into deep water, and the stern was left exposed to the force of the raging seas. It became evident that she was breaking up. It was just a question of time, a very short time at that, before the whole ship would disintegrate and wash ashore. And the last means of abandoning her had been destroyed with the lifeboat, the launching of which would have been extremely hazardous, if not downright suicidal. Help had to be summoned by radio.

Soon after the *Childar*'s SOS broke the air a few minutes after 7 A.M., the Coast Guard cutter *Redwing*, moored at her dock in Astoria, Oregon, prepared to go to her assistance. The liberty party was recalled, steam ordered, and the vessel made ready for sea. The Coast Guard surf stations, Point Adams Station and Cape Disappointment Station, were advised of the disaster and directed to send lifeboats to the scene of action. Then, at 8:30 A.M., the *Redwing* stood down the Columbia River with all possible speed.

A few minutes later, there came another message from the *Childar*, stating that she had no way to land the crew as her boats had been washed away, and it was feared that the ship was breaking up. Immediate assistance was necessary if either the ship or the crew were to be saved.

The *Redwing* answered this message but received no reply from the distressed vessel. The cutter then broadcast a message saying that she was proceeding with all possible speed and would arrive around 11 A.M.

The wind had shifted to a southwest gale and was increasing in velocity when the *Redwing* got under way. A tremendous sea had worked up, accompanied by squalls of rain. Lieutenant A. W. Davis, in temporary command of the *Redwing*, knew that extreme caution would be needed in crossing the bar, or the cutter herself might be driven upon the beach. Reaching the bar, it was found that heavy seas were breaking all the way across it—a dangerous condition, greatly aggravated by a strong ebb tide, plus the river current, which was at freshet height at this time.

Visibility was zero, and this, together with the fact that there was a phenomenal set toward the shoals on the north side of the bar, made the situation a precarious one for the Coast Guard cutter. Only superb seamanship would count now, for the shallow water and murky weather made it impossible to distinguish the breakers to shoreward. A wide berth was given the shoals, and the cutter fought her way through the rolling seas. At 10:30 A.M., the shoals were cleared, and as the weather lifted momentarily, the *Childar* was sighted ashore on North Spit with immense breakers passing entirely over her. The weather shut in almost immediately, and as difficulty was encountered by the *Redwing* in feeling her way around the breakers on North Spit, it was found necessary to approach the *Childar*'s position from seaward.

A few minutes after 11 A.M., the *Childar* was again sighted. The hapless vessel was a mass of wreckage above decks—masts gone, part of her superstructure carried away, her rails twisted and bent. There was a jagged hole in her starboard side, which, fortunately, was well above the water line.

From the bridge of the cutter, Lieutenant Davis studied the situation. He knew that it would be a risky job for the cutter to go close inshore and attempt to take the men off. The force of the seas would drive the Coast Guard vessel into the breakers before the rescue could be accomplished. Also, there was a bare possibility of the *Childar* herself being saved, providing she could be drawn away from the rocks and towed to sea. Once in open water, the removal of her crew would be a less dangerous job.

Accordingly, the Coast Guard officer brought his ship around into position and directed that a line be fired across to the disabled vessel. A steel cable was passed to the *Childar*, but she had no steam on her winches, and as the cable was too heavy to be handled by manpower alone, it was withdrawn in favor of a twelve-inch hawser.

By this time, the *Redwing* was dangerously close to the surf, and it was necessary to circle to seaward before firing the second line. On the way back, fog shut down again, making the task doubly difficult. Lieutenant Davis eased his vessel in toward shore, with the roar of the breakers beating against the *Childar*, sounding near at hand. Then, through the fog, the grounded ship loomed up, and the Hall line-throwing gun was fired, the line falling across her bow. The twelve-inch hawser was speedily drawn in and secured by the *Childar*'s crew.

The *Redwing* immediately took strain on the hawser, and the *Childar*, pulling clear of the shoals, wallowed out into deep water with a considerable list to starboard. As she appeared to be in a sinking condition, Lieutenant Davis dispatched a message to her captain, asking for his opinion on the ship's condition. Matthisen replied that the vessel might possibly stand a tow to port but that several of his injured men should be removed at once for immediate hospitalization.

The lifeboats from the Coast Guard surf stations had by this time reported for duty, and they were directed by the *Redwing*'s commander to remove the injured men from the *Childar*. Incidentally, there had been a casualty aboard the Cape Disappointment lifeboat while she was proceeding from her station to the scene of the distress. Chief Boatswain's Mate Lee Woodworth, in charge of the lifeboat, had been thrown against the engine room bulkhead by a big breaker that had boarded the Coast Guard boat. He suffered several broken ribs and was in terrible pain. Yet he kept on his feet with dogged courage—there was work to be done, and there could be no rest until his assignment had been completed.

It was decided that the Point Adams boat would remove the injured seamen while the lifeboat from Cape Disappointment Station stood by as a matter of precaution. Now had come a test for the seamanship of the Point Adams men, and they knew that the slightest mistake would add their lives to those already claimed by the sea.

The lifeboat headed under the *Childar*'s rail. Foot by foot she crept forward, riding the crest of the waves, then plunging into the trough between them. The men on the deck of the Norwegian vessel watched breathlessly. They were witnessing Coast Guard seamanship—a combination of iron nerve and unerring judgment.

Finally, the Coast Guard boat was alongside the *Childar*. Bracing themselves, two Coast Guardsmen prepared to receive the injured seamen as they were lowered over the side in improvised stretchers. Knowing that, at any moment, a sea might wash them overboard, they stayed at the job, those two surfmen, until the injured men were safely aboard and the lifeboat headed away from the *Childar*.

Lieutenant Davis later wrote in his report that the work of the two lifeboats had been superb; he gave unstinting praise to the Coast Guardsmen who manned them. He said: ". . . the assistance afforded by these stations was invaluable, and each displayed outstanding courage and knowledge of handling their boats and cooperated with this vessel to the fullest extent." Praise like that from a superior officer must be earned in the Coast Guard.

By this time, the *Redwing* had the *Childar* safely away from the shoals, but the work had just begun. Lieutenant Davis was determined,

if it was at all possible, to save her. Realizing that it would be impossible
to tow the disabled vessel back into the Columbia River—as an entrance
across the bar could never be made while the seas were running as high as
they were now—he decided to take her into Puget Sound.

At 1:30 P.M., the *Childar*'s bitts tore out due to the weakened condition of her plates, and the tow line was nearly lost. Instantly, the cutter's
speed was reduced and the bitts caught in the bow chocks, thus saving
the hawser from slipping into the sea. Difficulty was experienced in finding a means of making the hawser fast again, as the forecastlehead had
been swept clean. Finally, however, the tow line was made fast to the
anchor chains, and the tow was resumed.

The difficulties attending that towing job were almost insurmount-
able. The *Childar* had lost her rudder while on the rocks, and now she
yawed widely behind the *Redwing*, turning broadside and offering much
resistance to the sea. Speed was further reduced, for it was feared that
the hawser would part, and if that happened, she would inevitably drive
on the rocks again. The weather was steadily growing worse. At 8 P.M.,
a message was received from the *Childar*, requesting that the crew be
removed as the ship was giving signs of breaking up.

Lieutenant Davis released a message to Gray's Harbor surf station,
which was nearby, directing that a lifeboat be dispatched to remove the
rest of the *Childar*'s crew. At 11 P.M., the lifeboat arrived and, while the
Redwing trained her searchlights on the Norwegian vessel, accomplished
the extremely hazardous job of taking off eighteen persons without injuring a man. Captain Matthisen, the radioman, and three others of the
crew chose to stay aboard and trust to luck and the Coast Guard to pull
them and their vessel safely into port.

After putting the eighteen men aboard the *Redwing*, the lifeboat was
ordered to stand by the disabled vessel in order to remove the five men
in the event that the ship began to founder. The tow was resumed up the
coast, the *Childar* towing sluggishly with a tendency to sheer to port,
making it necessary to watch her very closely on some of the larger swells
in order to ease the strain on the hawser. Lieutenant Davis knew that,
should the line part now, there would be no chance of getting another
across due to the depletion of the *Childar*'s crew.

The task of towing her seemed almost hopeless. She appeared ready to break up at any minute. Unceasing vigilance and constant jockeying on the part of the *Redwing* were necessary to prevent the heavy swells from smashing down upon the *Childar* and tearing her away into the darkness. But somehow, the Coast Guardsmen nursed her through the night. When dawn broke, they found that the sea had moderated slightly and that the *Childar* seemed to be riding a trifle easier.

At 11 A.M., the cutter *Chelan* arrived and took up convoy with the *Redwing* and her tow. The weather would not permit the *Chelan* to relieve her sister cutter on the job, even had there been some way of getting another hawser aboard the *Childar*. Matthisen now reported that the vessel was filling with water and doubted if it would be possible to get her safely into port. But Lieutenant Davis stubbornly refused to admit defeat. As the sea had moderated, he felt that the chances of a successful tow were greatly enhanced. Moreover, the Straits of San Juan de Fuca were not far distant, and once the *Redwing* was in the comparatively calm waters of the Puget Sound, the fight would be over.

So the cutter kept at her tow throughout the day and far into the night. The weather moderated further, and around midnight, a message was received from the *Childar* saying:

GOOD FOR YOU. THINK WE'LL MAKE IT NOW.

When that message was received, the lifeboat from Gray's Harbor Station was released and directed to return to its base. The *Chelan* closed in on the *Childar* and kept a sharp lookout through the rest of the night. Then, at 8 A.M., the *Redwing* towed the disabled vessel into the Straits of San Juan de Fuca.

Here, as the *Redwing's* commander had known, the water was comparatively smooth, being protected from the open sea by a stretching finger of land, and there was no further danger of the *Childar* being swamped by huge seas. The ship was considerably down by the stern, and an inspection showed that number four hold was flooded, number five partially so, with water to the top of the tanks in the engine room. It seemed incredible that she could have remained afloat an hour, miraculous that she was towed 205 miles in such rough weather.

Off Port Angeles, Washington, that afternoon, the *Childar* and her crew were turned over to the commercial tug *Roosevelt*. The underwriters of the ship had deemed it necessary that the *Childar* be towed to the nearest docking and repair facilities, and accordingly, she was towed to Victoria, British Columbia. The *Chelan* returned to her regular station, the *Redwing* proceeding to Port Angeles, where she docked for the night so that her officers and men could get the first sustained rest they had had for four days. Thus, another job by the Coast Guard, in which twenty-six persons were actually snatched from death and in which a vessel and cargo worth over $700,000 had been saved, was finished satisfactorily and competently.

Charles Nichols loved the sea. Otherwise, he would never have quit the peaceful business of raising chickens to return to salt water after a number of years of absence. Funny thing about the sea. While you are on it, and when it starts kicking up, you curse it with every breath, but when you're away, it's like a separation from a loved one.

Charles Nichols found this so and, finally, came back to the life he loved—and hated.

On February 7, 1934, Nichols, master of the schooner *Purnell T. White*, found himself somewhere off the coast of North Carolina, his vessel bereft of sails and one of her pumps disabled—the result of nearly a week of stormy weather. He was flying distress signals, for the Atlantic Ocean in February is no place for a sailing vessel to be without sails, and Nichols knew it.

Sighting the *White*'s signals, the SS *Maiden Creek* steamed close and spoke to the schooner. Nichols explained the circumstances and asked that the Coast Guard be notified. The steamer at once sent out a call for the Coast Guard, and when the cutter *Mendota*, which was on patrol off the Virginia coast, answered, the *Maiden Creek* agreed to stand by the *White* until the *Mendota* arrived.

The 125-foot Coast Guard patrol boat *Tiger* was nearer to the two vessels and was directed by the *Mendota* to proceed and stand by for the cutter's arrival. The *Tiger* reached the scene about noon and found that the *White*, despite her loss of sails, appeared to be in good condition. The steamer continued on her course, being informed by the *Tiger* that she could be of no further assistance.

The patrol boat transmitted radio signals and guided the *Mendota* to the spot. Later that afternoon, the cutter, under the command of Commander Keaster, immediately took steps to tow the *White* into safe waters. The weather at this time was favorable, and no difficulties were anticipated in towing the schooner into Lynnhaven Roads.

A twelve-inch towing hawser was passed to the *White*, and then the *Mendota* got under way. The *Tiger* acted as convoy until the early morning of February 8, when she was dispatched by the *Mendota* to investigate reports of a barge being ashore further down the coast.

Toward evening, difficulties were encountered in the shape of a hard northeasterly, and Commander Keaster found it necessary to reduce speed until only steerageway was made by the cutter and her tow. The *Mendota* attempted to shift course so that the force of the sea would not cause the *White*'s seams to open, and this effort was partially successful. However, on the morning of February 9, a course to seaward had to be taken in order to avoid being carried upon a lee shore below False Cape, Virginia.

By this time the wind had gradually shifted to the north and north-northwest and had increased to a fresh gale, accompanied by a snow blizzard and heavy vapor, with resulting low visibility. The seas were following the wind and had become very rough. The *White* was beginning to roll and pitch heavily, and Commander Keaster, considering the possibility of her cargo of lumber shifting, ordered an oil slick made for the schooner. (An oil slick is made by releasing fuel oil on the water in order to smooth the seas down with its weight.) The slick in this case, appeared to be of little help, for about an hour later, the *White* hoisted an international signal to the effect that she was waterlogged. The whirling snow and wispy vapor hid the schooner from sight most of the time, but extra lookouts were posted to be on the alert for further distress signals.

During the late afternoon, the lookouts noticed that the *White* was beginning to roll more irregularly, more to port than to starboard. There were, however, no signals to the effect that the crew wished to abandon her; a red flare had been agreed upon as an indication that they wished to be removed. While no such signal was made, Commander Keaster realized that the schooner's crew would have to be taken aboard without further delay.

There were two possible means of accomplishing this. One was to put a boat over the side from the *Mendota*—an idea that Keaster almost instantly abandoned. The heavy seas and poor visibility made such an attempt too risky, for it was extremely likely that the boat would be swept away and irretrievably lost in the leeward wrack. The second and only feasible means of rescue would entail jockeying the *Mendota* up to the schooner's side to allow the crew to catch lines by which they could be taken aboard the cutter.

Commander Keaster realized that it would be a most dangerous undertaking. The schooner might easily be sunk if the *Mendota* was thrown against her by a sea; indeed, the cutter herself would be damaged in such a collision. There was no time for hesitancy, however, and Commander Keaster ordered that preparations be made aboard the *Mendota*. Boatswain's pipes shrilled. Coast Guardsmen, clad in oilskins and boots, brought out heaving lines, life jackets, rafts, and a line-throwing gun. Pilot ladders were lowered over the bow. Sand was sprinkled around ice on the desks to afford surer footing.

A signal was flashed to the *White* directing her to throw off the hawser, but evidently, it was not possible for any of the schooner's crew to get forward to cut the *Mendota*'s line. A hundred-pound weight was hung on the hawser near the *Mendota*'s stern, and as the cutter went forward at full speed, the line was cut. Commander Keaster was taking no chances of the hawser fouling the *Mendota*'s screw.

The cutter then circled to approach the stern of the schooner, but the *White* had been lost to sight in a blanket of snow and darkness. Searchlights probed for her, and when she was finally sighted again, men were seen huddled under the shelter of the deckhouse.

Seas were breaking over her, and ice had formed on the rails. She had taken a pronounced list to port, caused, doubtlessly, by shifting cargo. The end was near.

The *Mendota* managed to approach within thirty feet of the schooner's port quarter, and after several attempts, a heaving line was thrown to her. The men on the *White*'s deck grabbed the line and drew it in eagerly. The line was followed by a three-inch cable to which a life raft had been made fast.

Having lashed himself to the raft, Leon C. Spence, a seaman from the *White*, was drawn across to the cutter by willing hands. Suffering from exposure and badly frozen hands, he was taken below and given first aid by the chief pharmacist's mate.

The raft was sent back to the *White*, and this time, John Olsen, another seaman, and a cook known as Jim went aboard the raft and prepared themselves for the hazardous trip through the slashing seas. They were drawn up as closely as possible to the ladders on the *Mendota*'s bow and lines were let down to them. These the men tried to fasten about themselves, but it soon became apparent from their actions that they had been greatly weakened by exposure. Their efforts were clumsy and feeble, hampered by frozen hands.

A Coast Guardsman started down the ladder to help the two men, a dangerous thing to do with the vessel pitching and bouncing in the water. But he was too late. Panic-stricken, they made another desperate effort to reach the ladders. In so doing, they lost their footing on the ice-covered raft and were washed overboard, never to be seen again.

By this time, the *Mendota*'s bow was slightly ahead of the *White*. The cutter persisted in sailing down upon her. Then, suddenly, the inch line between the two vessels parted, and the cutter drifted to leeward so fast it became necessary to maneuver into position again.

Commander Keaster backed the *Mendota* into the wind and sea so that sight of the schooner might not be lost at this critical time. Then final disaster struck the *White*. She began to keel over rapidly and, a few moments later, lay on her beam ends with the four masts and rigging in the water on the port side. The four men remaining on the schooner managed to scramble for the icy rail, clinging to it as she keeled over. Two were aft near the stern, the other two forward about forty feet away.

Commander Keaster realized that the four men were in an extremely perilous position, in imminent danger of being swept away. He decided now that the situation warranted the use of any method, no matter how risky, to affect a rescue. Driving the *Mendota* forward, he pushed her stern into the port quarter bilge of the schooner between the two groups of men and allowed the stern to ride over the keel. Lines were thrown to the four men, and having lashed themselves together, the pair on the forward rail were pulled aboard the *Mendota*. After futile attempts to keep his hold, one of the two men left on the wreck was washed overboard and vanished. The remaining man was Charles Nichols, who had left a chicken farm to return to sea.

Several attempts to get lines to him failed, some falling almost within reach, but Nichols apparently did not wish to risk the loss of his precarious hold on the schooner's rail. Finally, a line fell directly across him, and he fastened it about his body. Waving to the men on the cutter's deck, he jumped into the water.

The sea claimed Charles Nichols. The line that would have saved him fouled the schooner's keel. A huge breaker crashed down on the unfortunate man, driving him under the submerged rail. The line went taut, then as suddenly slackened and drifted clear. Nichols was gone.

It was now imperative to save the *Mendota* from possible damage. She had worked herself almost on top of the *White*, and the hull was bumping hard against the schooner's keel. By using full left rudder and going ahead on the engines, it was possible to push the hull around somewhat to windward. Thus, the *Mendota* managed to pull clear. Sweeping the searchlights around in all directions, the cutter tried unsuccessfully to find the missing men. Indeed, after clearing the hull of the *White*, the schooner, despite sharp lookouts and the searchlights, was almost instantly lost to view. Commander Keaster was reluctant to leave as long as there was a possible chance of finding the men, but after an hour's cruising, he headed the cutter toward Norfolk. The three rescued men needed hospitalization, having suffered frozen limbs from exposure.

Commander Keaster's official report to headquarters stated, ". . . in considering this case, particular attention must be given to the circumstances existing during the rescue operations. Action by the *Mendota*'s crew and the survivors was handicapped by the severe cold, the heavy seas, and blinding snow and vapor. Of the seven men of the *White*'s crew, six of these men were given lines to come aboard, and the fact that only three of them were saved was a grave disappointment to all who did their utmost to bring them aboard."

The rescue of the three men from the *White* was a brilliant feat. Another Coast Guard victory, partial though it was, wrested bravely from the sea.

Chapter VII

DEATH GOES TO SEA

The Coast Guardsmen like their work; they like to see it well done and will bend every effort to see that it is well done. Such jobs as the *Childar* and the *Purnell T. White* always bring a comfortable feeling of a task well done to the heart of the Coast Guardsman. But there are other incidents, incidents of which the Coast Guard does not like to speak, that bring a tightness to the lips of the service man and a grim look to his eye. Such an incident was the killing of two Coast Guardsmen and a federal agent aboard the CG 249 on August 7, 1927.

The passage of the Eighteenth Amendment to the Constitution of the United States plunged the Coast Guard into a bitter war that was to last for nearly fourteen years. Prohibition made the smuggling of liquor into the United States a profitable venture, a lure for desperate characters, men fascinated by the prospect of adventure and easy money—big money!

At one time there was an estimated fleet of 250 foreign vessels engaged in the business of smuggling liquor into the United States. The Coast Guard was given orders to stop this smuggling. It was a big job, calling for long, almost endless patrols in all kinds of weather. In calm blue water, through towering seas, and in raging winter gales and fog, the patrols went on against the small, rakish craft that employed cunning ruses to land their cargoes. Poison gas, smoke screens, and machine guns were the smugglers' common weapons.

Men died.

Sometimes they were smugglers, defying the laws of the United States; all too often, they were Coast Guardsmen, dying in battle as men of the Coast Guard had died in every war this country has known. The motto of the Service, Semper Paratus—Always Ready—is enhanced and exemplified by the facts now to be presented.

It was a pleasant Sunday morning when the liberty party returned to the Coast Guard base at Fort Lauderdale, Florida. They came in through the gate—healthy looking, bronze-faced men, clad in white. The sentry on watch saluted briskly as Boatswain Sidney C. Sanderlin, tall and sharp-eyed, entered the reservation.

The sentry said: "The commanding officer wished to see you, sir. You are to report to him immediately."

Boatswain Sanderlin nodded cheerily and went on. He found Lieutenant Becksith Jordon, commander of the base, standing on the dock, talking to a big man in gray civilian clothes. Returning Sanderlin's salute, Lieutenant Jordon introduced him to Robert Webster, secret service operative.

Jordon then explained that Mr. Webster had orders from his department to proceed to Bimini, British West Indies, to confer with schooner owners, who had important information concerning a recent flood of counterfeit money that had been issuing from the islands. The Coast Guard, Jordon added, was to furnish transportation, and Sanderlin was to carry out the mission of the CG 249, a trim seventy-five-foot patrol boat.

It was later that same morning when Sanderlin, secret service agent aboard, headed the 249 out to sea, a rendezvous with death!

The crew on that memorable trip consisted of Boatswain Sidney C. Sanderlin, in command of the CG 249; Boatswain's Mates Lawrence F. Tuten and John A. Robinson; Motor Machinist's Mates Frank Lehman and Victor A. Lamby; Seamen H. M. Caudle and Jodie L. Hollingsworth. In addition to the crew, as passenger, was the secret service operative, Robert Webster.

A bright morning sun glinting on the dancing, choppy water gave no hint of what was to follow. The 249 was about forty miles out from the base when a small schooner was sighted approaching from the direction of Bimini. In those days, Bimini was the base for the liquor boats, and it was from there that the "rummies" made their runs across to the Florida coast with cargoes of alcohol.

Boatswain Sanderlin leveled powerful glasses on the schooner. About forty feet long, she was apparently powered with auxiliary motors, and the numbers V13997 were painted on her bow. She turned slightly in toward the shore when, at Sanderlin's command, the CG 249 swung toward her.

"A rummy," the boatswain said, laconically; and he reached for the signal cord.

The blast sounded, but the schooner paid no heed to this signal. She plowed on, heading closer in toward land.

Sanderlin next ordered a blank shot fired from the one-pounder on deck. Twice this was done without result. "All right," he snapped. "Fire a shot across her bow!"

The solid shell screamed into the wind and went howling out in front of the schooner. She hove to instantly and lay rocking in the swells while the Coast Guard circled her and came up on her port side. When two vessels had been lashed together, Boatswain Sanderlin jumped onto the deck of the schooner, gun in hand. A tall, bearded man came out of her pilot house; he was about fifty, with small, snake-like eyes.

Sanderlin searched him but found no weapons. "Don't you know the regulations about stopping when signaled by a Coast Guard vessel?" he demanded. "Let's see your papers."

Sanderlin watched the man closely as he went into the pilot house and presently returned. The papers he showed to the Coast Guardsman indicated that the schooner belonged to James Horace Alderman of Miami.

Sanderlin nodded. "You're Alderman, I suppose? What're you carrying on this boat?"

Alderman spat carefully over the side. "Nothing much. Fishing gear. Going to do a little fishing." His eyes darted about nervously, then focused on the deck at his feet.

Grinning tightly to himself, Sanderlin went aft to the hold; he had seen many rum runners in his day, and he felt sure he was not mistaken in this man.

He was right. In the hold, he found twenty cases of liquor in burlap bags. He poked his head up through the hatch and called for his chief boatswain's mate Tuten. Assisted by the latter and Seaman Hollingsworth, he brought the liquor up and placed it on the schooner's deck while Alderman watched with a glint in his sullen eyes.

"Tough luck, fellow," Sanderlin said to him. "You're caught with the goods."

It should be understood that the Coast Guard bore no personal grudge against smugglers as men but merely endeavored to carry out orders to the best of its ability. When caught, smugglers usually grinned ruefully and shrugged. They knew that the syndicates which employed them would soon buy them out and that, with luck, they'd be back working at the same old stand. Alderman, however, hated the Coast Guard, hated it with all the power of his warped soul. His friends had been killed by the Coast Guard. It didn't matter to him that they had died resisting lawful authority and that the Guardsmen had killed in self-defense. His hate was instilled in his heart; blood alone could erase it.

While Sanderlin herded Alderman across to the patrol boat, another man named Weech—a companion and henchman of Alderman—was left on board the rummy boat with instructions to make ready to be towed back to the Coast Guard base.

Boatswain Sanderlin went forward to the pilot house in a thoughtful mood. He was in a predicament. He had orders to carry the secret service man into Bimini, but one of the Coast Guard's primary duties was its campaign against such persons as Alderman. Picking up the radio microphone, he switched on the transmitter and prepared to call the radio station at base six for advice. He glanced back at Alderman, who was standing just outside the door, a queer expression on his dirty face. Webster, the secret service man, leaned against the rail, staring at the rummy boat. Victor Lamby, the motor machinist's mate, was coming up out of the forecastle's hatch, a grin on his homely face. That grin was the last thing Boatswain Sanderlin was conscious of in life. Suddenly, Alderman jerked out a .45 caliber automatic and shot the Coast Guard officer in the back of the neck. Blood spewing from his wound, Sanderlin fell across the desk; he tumbled to the deck, dead.

There has always been a mystery attached to Alderman's possession of the gun. He had been thoroughly searched and watched carefully; yet, somehow, with the cunning of a trapped animal, he had managed to get his hands on a weapon. The only logical explanation is that Alderman had placed a gun at some convenient place on deck, and as he went by on his way across to the patrol boat, he had slipped it into his shirt.

There was no time for conjecture now. With the report, Lamby spun round in time to see his commanding officer fall. Shouting, he started to run aft, intending to get to the armory, where there were plenty of weapons. But Alderman hurdled Sanderlin's body and, rushing through the pilot house, fired at Lamby just as the Coast Guardsman was turning the corner. The bullet struck the boy's side and tore upward to rest in his spine. He fell to the deck, his lower limbs paralyzed and useless. He kept crawling, trying to escape the death that was just behind him. He pulled his body up over the engine room hatch coaming, fell ten feet to the floor plates, and lay still—bloody but still faintly breathing.

Joe Robinson, another member of the Coast Guard crew, was standing near the engine room hatch, a heavy wrench in his hand. When he saw what had happened to Lamby, he hurled the wrench at the killer's head. Missed! The man whirled around, snarling, and brought his gun down on Robinson, who promptly dove overboard.

Alderman then herded the remaining Coast Guardsmen and the secret service agent back on the stern of the rummy boat. "I've got two of you damn Coast Guardsmen," he snarled. "I'll get the whole bunch of you before this thing's over!" He called his henchman Weech and asked: "You with me in this thing, Weech?"

Weech nodded, his eyes on the gun in the man's hand.

"Okay," said Alderman. "Go down in the engine room of the patrol boat. Break all the gas lines and set fire to the damn thing. I'll bump these guys and throw the bodies over on the deck."

Cold sweat broke out on the faces of the prisoners. The secret service agent started talking fast: "What the hell are you trying to do? You can't get away with this! You don't want to have all this blood on your hands just for the sake of a few sacks of liquor, do you? Put us over the side in a small boat, and let us take our chances that way."

"Give you a chance, hell!" Alderman laughed harshly. "The only chance I'm going to give you is a chance to say your prayers! Get down on your knees if you want to make your peace!"

Meanwhile, the shriveled-faced Weech had clambered down into the engine room of the patrol boat, where he found Lamby sprawled out on the floor plates, blood welling from his mouth.

"Get up, damn you," Weech said, pulling the Coast Guardsman to a sitting position and staring into his wan face. "You're going to help me flood this place with gasoline. We're going to burn her."

"I'm paralyzed—dying," Lamby groaned. "I—"

Weech kicked him in the side, the side so recently pierced by a bullet. Almost mad with pain, Lamby reached up and handed Weech a wrench. Then he dropped back to the deck, unconscious.

Going to the pipe lines, Weech immediately tore them down, letting the gasoline pour into the bilges. This done, he climbed back up on deck and sang out to Alderman: "There's a wounded man in the engine room. What'll I do with him?"

Alderman cursed. "Kill the—!"

Weech scratched his head. "But I haven't a gun. You never gave me a gun!"

"Damn it, then, set fire to the gasoline. Burn him alive. We're wasting time!"

Chief Boatswain's Mate Tuten said: "You're going to blow us all up. If you set fire to the patrol boat, there'll be an explosion, and your schooner will catch fire. We won't have a chance." He was stalling for time, a chance to rush the killer. But Alderman now held Sanderlin's gun, as well as his own. His eyes gleamed, and Tuten thought that he was going to shoot.

Instead, Alderman cursed and said: "I don't need any help from you stinking Coast Guard. You've caused me enough trouble already." Tuten's suggestion, however, seemed to find favor with him, for he sent Weech to start up the motors of the schooner.

Meanwhile, Joe Robinson, the man who had jumped over the side to avoid being shot by Alderman, pulled himself up over the rail. Alderman watched him with baleful eyes but made no belligerent move as Robinson joined his companions.

The Coast Guardsmen knew that as soon as the motors of the schooner were started, Alderman would kill them and set fire to the patrol boat. His scheme was crude but evidently destined to succeed. Sanderlin had failed to contact the base via radio. When the patrol boat failed to show up at her destination, her disappearance would be charged to fire and explosion. Alderman would never be suspected, much less apprehended. It is more than probable that Alderman, in order to be entirely in the clear, was also plotting Weech's death.

Weech was having trouble with the motors. He got them started, only to have them sputter and die. Alderman sat on a coil of rope and shouted instructions and curses at the monkey-faced little smuggler.

Suddenly, Robinson saw a small ice pick lying on the deck just behind him. Lucky! he thought. While the others screened him, he stooped quickly and picked it up. It was short, almost too short, but it was a weapon. There came a backfire from the schooner's motors. Alderman jerked his head around in alarm. Through the minds of the tense men before him ran the same thought—NOW!

Hollingsworth and Webster lunged forward, closely followed by the rest of the Coast Guardsmen. Webster grappled with the killer and got his hand on the gun in the man's right hand. But he missed the other gun. Alderman jerked the .45 up. It belched almost directly into the face of Webster, who fell to the deck, dead.

Hollingsworth was fighting, fighting for his life and for the lives of the others. He had almost succeeded in disarming Alderman when the man shot him through the arm, then again, through the right eye. Reeling to the rail, he fell over into the shark-infested water.

Robinson was close in now, striking, stabbing Alderman with the ice pick, stabbing for a vital spot. The smuggler seemed possessed of the lives of a thousand devils. He bellowed and cursed and struggled. Then, finally, it was all over. Alderman went down, bleeding, under the combined assault of the Coast Guardsmen, who beat him into insensibility.

Seaman Caudle picked Alderman's gun from the deck. "There's another one of these rats down in the engine room," he said grimly. "I'll get him." Letting himself down through the engine room hatch, he came upon Weech huddled behind the engines. The little smuggler saw the look on Caudle's face and, picking up a heavy pipe, prepared to fight.

Caudle jabbed the gun at the man and pulled the trigger. The gun failed to fire, presumably damaged when it had fallen from Alderman's hand. The Coast Guardsman threw it at Weech and dived into him, fists flying. He fought the man off his feet, dragged him up the ladder, and threw him on deck.

It was Caudle and Robinson who fished Hollingsworth out of the water. The badly wounded boy had been hanging on desperately to a line swaying over the side of the patrol boat. His right eye was gone and he was bleeding terribly from the mouth. As they brought him over the side, he grinned faintly and tried to thank his shipmates. They put him down on the deck and made him as comfortable as possible, turning his head so that the blood might run freely from his mouth.

They went down into the engine room and hoisted Lamby out. The machinist's mate was badly hurt and in terrible pain. He kept saying over and over again: "I couldn't get to the armory! I couldn't get to the armory!"

There was nothing to be done for secret service agent Webster and Boatswain Sanderlin. They were dead. Tuten assumed command of the vessel and radioed the base, giving full details and asking for immediate assistance. The Coast Guard base, under the command of Lieutenant Jordon, swung into action. Ensign Hahn, in a fast speedboat, was directed to the scene with six petty officers and two seamen. All available boats at sea were ordered to proceed to the assistance of the CG 249.

Ensign Hahn arrived in a couple of hours, and the wounded men and the bodies of the dead were loaded onto the speedboat for the run back to the base. The officer left orders with a sister ship of the CG 249 that the 249 and the rummy boat be repaired and brought into the base. Alderman and his helper, Weech, were shackled together in the hold of the patrol boat.

The Coast Guardsmen went to work repairing the damaged engine room of the 249. They pumped the bilges free of gasoline, and presently, the craft was able to move under its own power. But they were not so lucky with the rummy boat. In trying to start the motors, there was a backfire and an explosion; the vessel caught fire, burning to the water's edge and finally sinking.

Once back at the base, the murderer and his henchman were turned over to the sheriff for safekeeping. The peace officer deemed it necessary that the two men be removed, and accordingly, they were carried to the

Miami jail. Then began the long series of legal fights for which the un-
derworld of the Florida east coast banded together in a desperate attempt
to save Alderman's life.

For two years this went on; threats were made, witnesses approached.
The commander of the Coast Guard base received a letter saying that for
Alderman's life, three Coast Guardsmen would die. A man who enjoyed
a high reputation in Florida wrote a letter to the president of the United
States, in which he pointed out the faults of the Coast Guard and asked
that the Alderman case be investigated fully before action was taken.

The Coast Guard, however, considered further investigation unnecessary. Two of their men were dead, Lamby having died on the morning
of August 11, 1927. They had been killed in the performance of their
duty, and all the Coast Guard wanted now was to see their murderers
punished in accordance with the law.

The amount of the bail set in the cases of Alderman and Weech was
surprisingly small—$25,000. The Coast Guard protested, as did the se-
cret service, which had sent a man down to investigate Webster's death.
They said that such a sum as bail was unthinkable.

At last, the two men went on trial. Weech saw a way out. The star
witness against Alderman, he claimed that he had been forced to do Alderman's bidding. Weech was sentenced to a year and a day, which he
subsequently served in the Atlanta Federal Prison.

Alderman's defense was weak. He claimed that he had mistaken the
Coast Guardsmen for pirates and hijackers and was merely protecting his
vessel against their attacks. He did not explain how he had come to miss
the huge CG 249 painted in white on the gray bow of the patrol boat and
the Coast Guard ensign flying at the mast. The jury rejected his defense
plea and returned a verdict of guilty.

Alderman was sentenced to hang.

All pleas for mercy by the doomed man's family and friends were
denied. Then the problem of where Alderman was to be hanged arose.
He had committed murder on the high seas and therefore was a federal
responsibility. Judge Ritter ordered that he be hanged on the government
reservation at base six, Fort Lauderdale, Florida. The Coast Guard deeply
regretted the choice of this location.

Accordingly, Alderman, who had come to be known as the "lone wolf of the sea," was hanged on August 17, 1929, in the gray light of the dawn while sentries patrolled the grounds. The loud clap of the trap door as it fell from under Alderman's feet pronounced an end to the blood-spattered saga of the "Battle at Sea."

The Sinking of the Schooner I'm Alone

March 20, 1929. The Gulf of Mexico. Dirty, gray water whipped by a frenzied wind. Lowering clouds and drifting, misty rain. The Coast Guard patrol boat *Wolcott* was on its regular patrol, out from the base at Pascagoula, Mississippi. Boatswain Frank Paul, with a well-earned reputation of competency in the Coast Guard, commanded her. He stood in the pilot house, his legs spread apart, bracing himself against the pitch and roll of the ship. He was alert, as were the helmsman and the lookout atop the pilot house. All knew that this was "rummy" country, with dark, speedy boats lying out there in the rolling swells, waiting to make a dash for shore with their cargoes of liquor and aliens. It was the *Wolcott*'s job to see that those cargoes were never landed.

"Boat dead ahead, sir!" the lookout called down suddenly through the voice tube.

Five or ten minutes passed before the boat was visible to those in the pilot house. Then they saw her—a dark shape lying in the trough of the sea. Another ten minutes and Boatswain Paul could read the name on the stern of the rakish-looking craft.

She was the schooner *I'm Alone*—known to the entire Coast Guard patrol force operating in the Gulf, for she headed the lists of vessels suspected of smuggling. Though chased a number of times, she was never caught, always managing to get away in fog or darkness. This time, Boatswain Paul resolved grimly, she would not escape. Rapidly computing his position, he found that he was well within the twelve-mile limit. He logged that with satisfaction and, turning to the signal cord, jerked it hard. The horn bellowed hoarsely.

Immediately the schooner came to life. Her motors started, and she went ahead through the waves, spray bursting over her squatty pilot

house. Paul whistled softly. Plenty of speed there! He swung to the an-
nunciators and threw the handles down to full speed. The *Wolcott's* motors hummed and the powerful little patrol boat cut through the water.

Again, the *Wolcott's* signal to heave to went out, but the *I'm Alone*
kept on. The Coast Guard officer ordered his gun crew to their posts and
directed that the customary blank shot be fired.

WHAM!

The skipper of the *I'm Alone* paid no heed. By this time, the *Wolcott* had pulled up within hailing distance. Paul megaphoned across the
water: "HEAVE TO, THERE! THIS IS A COAST GUARD VESSEL!
HEAVE TO!"

A man came to the door of the *I'm Alone*, shook his head, and motioned derisively with his hands.

"Solid shot this time, across her bow!" snapped the officer.

The gun roared sharply. The shell screamed past the schooner, struck
the water a half mile away, and skidded out to sea.

This time the captain of the schooner came out on deck. He ran a flag
up on the mainmast of the schooner. It fluttered out in the wind, and the
Coast Guardsmen saw that it was the British flag. The man came to the
rail and motioned the Coast Guard patrol boat closer and shouted that
if the captain of the *Wolcott* would come across unarmed, he'd be glad to
talk over the situation.

Paul agreed. He stationed four men on deck with rifles and then, as
the cutter pulled up alongside the schooner, hopped across to the deck.

A tall, gray-haired man with a military bearing met him and introduced himself as Captain John T. Randall. The captain of the schooner
was frank and courteous. He admitted openly that he was carrying a load
of liquor, that he had come from British Honduras, and that although his
clearing papers showed that he was bound for Hamilton, Bermuda, he
had no intention of going there. In other words, his liquor was destined
for an American port, and he didn't care who knew it.

They went into the captain's room. Paul knew then that the whole
thing had been just a ruse to let him see that the skipper meant to fight
any boarding party that might come aboard his vessel. There was a rifle

rack in the captain's room with slots for eight rifles. There was only one rifle in the rack. A heavy automatic lay close at hand.

Paul said tightly: "There's nothing to be gained by talking, Captain. You're carrying liquor. It is my duty to stop you."

The tall man's face clouded quickly. "If you fire on this boat, you'll be firing on the British flag. You know what that means. I've been an officer in the British Navy. I'll be damned if I surrender to you!"

The Coast Guard officer said quietly: "Then I'll sink you, Captain."

Randall tried one more argument. "I'm outside the treaty boundary. I'm on the high seas."

Boatswain Paul smiled briefly. "When this chase started, you were within boundary. That's the principle of the 'hot chase.' It was originated during the Civil War, and every civilized country in the world recognizes its validity. I can chase you to Kingdom Come and back and still be within my rights and authority!"

The schooner's captain scowled darkly. "If you board me, there'll be a fight. My men are armed. I'll see this vessel sunk before I'll allow her to be taken by you!"

Boatswain Paul saw that there was no need for further talking. He went across to the *Wolcott* and, turning once more to Randall, said: "I'll start firing in ten minutes!"

The captain of the *I'm Alone* shook his fist at the patrol boat. "Fire and be damned to you!" he shouted.

Ten minutes later, a shell from the patrol boat howled into the rigging of the schooner. Again, the demand to heave to came from Boatswain Paul. And again, it was refused.

Then, on the third shot, the three-inch gun jammed, went out of commission. The officer called for volunteers, and the entire ship's company, those who could be spared from necessary watches, responded. Paul told them that he was going to board the *I'm Alone* and that he wanted them to know that the smugglers were armed and meant to fight. Some of the Coast Guardsmen would not come back, he said. He was silent for a moment, waiting for the volunteers to back down, while in his heart, he knew they would not. Not one asked to be released.

Boatswain Paul shook his head. He thought a lot of these men of his. It was too great a price to pay. There was another way, a way in which Coast Guardsmen would not die. Abruptly, he dismissed the boarding party and went up into the radio room.

In a few minutes, the radio operator had contacted the base and was giving the message. Boatswain Paul explained about the three-inch gun, the dangers attending upon the boarding of the *I'm Alone*, and asked that another patrol boat contact the *Wolcott* and her trail. The message was passed to Lieutenant Commander A. H. Bixby, the commander of the base.

At 5:30 P.M., the *Dexter*, a sister ship of the *Wolcott*, under the command of Boatswain A. W. Powell, put to sea, heading out to intercept the *I'm Alone*. Then, at 7:30 A.M., March 22, the *Dexter* made contact with the *Wolcott*. She came within hailing range, and the two Coast Guard officers discussed the situation.

Paul explained that the schooner had repeatedly ignored the command to heave to for boarding and that he had been trailing the *I'm Alone* since the preceding Wednesday morning. He further stated that the three-inch gun on the *Wolcott* was out of commission and that he did not have another gun of sufficient caliber to stop the *I'm Alone*.

Powell arranged then with the captain of the *Wolcott* that his cutter would make the necessary show of force and that the crew of the *Wolcott* would do the boarding. Then he immediately hailed the skipper of the rummy schooner and demanded that he stop his vessel at once. The man laughed and pointed at the British flag flying from the mast head.

At about 8:22 A.M., Powell ordered solid shot fired into the rigging of the schooner. He took great pains to instruct his men to fire forward of the main mast, as the crew of the rummy boat seemed to be gathered on the stern. He also directed that machine gun fire be played on the *I'm Alone's* hull, in order that a slow leak might be brought about in the vessel.

At 8:30 A.M., after a number of shells had struck the schooner, the Coast Guard officer ordered the firing held and again demanded that the schooner surrender. This demand, as were the others, was refused.

Accordingly, it being apparent that the skipper of the *I'm Alone* had no intention of stopping his vessel for a search, Powell directed that the three-inch gun be trained on the schooner's hull below the water line. At 8:50 A.M., a three-inch shell struck the schooner below the water line, just forward of the main mast, and tore a great hole in her side.

Immediately, the vessel started filling with water, settling quickly by the head. At 9:03 A.M., she turned over and sank, her crew and captain having jumped just before she capsized.

A boat was put over immediately from the *Dexter* and the *Wolcott* in an effort to rescue the crew of the schooner. Charles B. Raeburn, seaman first class aboard the *Dexter*, saw a man floating in an unconscious state just below the surface of the water. The seaman jumped over the side into the rough sea and brought the man to the surface, supporting him above the water until both were picked up by a boat from the *Wolcott*.

Four other men, including the captain of the schooner, were picked up by means of lines and brought aboard the *Dexter*. The remainder of the crew had been rescued by the *Wolcott*.

The two patrol boats cruised around searching for wreckage and ascertaining that all the crew of the *I'm Alone* had been rescued. The man that Seaman Raeburn had saved turned out to be the first mate of the *I'm Alone* and was apparently drowned. Artificial respiration was practiced on him steadily for three hours without success.

A cousin of the drowned man, another member of the crew of the *I'm Alone*, swore under oath that he had gone to Captain Randall several times during the chase and pleaded with him to stop the schooner. The captain had sworn at him and said: "If you shut those engines off, I'll shoot you! I'm too damn proud of these medals I won in the war to surrender this vessel!"

The man, whose name was Chesley Hobbs, further stated that had it not been for the hardheadedness of the schooner's skipper, his cousin would not have died. There was, according to his sworn statements, not a single life preserver aboard the *I'm Alone*. Certainly, there were none found by the *Dexter* and the *Wolcott*, and none were used by the crew when they jumped overboard at the end.

The crew of the *I'm Alone* were taken to New Orleans and turned over to the United States Marshalls there. Immediately, notes of diplomatic representation were received from England. A British ship had been sunk; the British flag had been fired on by an armed vessel of the United States. It was a situation that was fraught with tense possibilities.

The United States held that the *I'm Alone* had been discovered well within the treaty limits. And this had to be proven. As luck would have it, at the beginning of the chase, the *Wolcott* and the *I'm Alone* had passed the American tanker *Hadnot*. With unusual intuition, Boatswain Paul had requested the skipper of the *Hadnot* to plot the position of the *Wolcott* and the *I'm Alone* and log it. Later, this was to prove important, for knowing the speed of the two vessels and the direction from which they came, it was possible to determine almost exactly the position of the *I'm Alone* at the start of the chase.

It was pointed out by the Coast Guard that the *I'm Alone* was a known smuggler. She had been given every chance to stop and submit to a search. If she had been clear, this opportunity would no doubt have been accepted. The captain, John T. Randall, had openly admitted that his vessel was carrying liquor; he had refused to allow a peaceful search.

Once, Randall had stood at the rail and shouted to the Coast Guard patrol boat: "Come across and get a little lead!"

It was proven conclusively that the mate had drowned when he jumped from the schooner. Had there been life preservers, it was safe to assume that he would not have died before help from the two Coast Guard boats reached him.

Correspondence went back and forth between the two countries. Gradually, the tension over the sinking of the *I'm Alone* lessened. As one high official in the Coast Guard headquarters at Washington said: "The Coast Guard rises or falls by this case. We are at constant war and must not, for one moment, relax our vigilance. We are entrusted with a great responsibility, and the Coast Guard will meet this responsibility firmly."

The Coast Guard has done, and will do, just that.

Chapter VIII

WINGS AND MEN

With the development of Coast Guard aviation, it is no longer possible for such smuggler ships as the *I'm Alone* and others to lay to out on the edge of the twelve-mile limit without being discovered. Coast Guard planes with cruising ranges of 500 to 1,500 miles sweep out to sea, plot the positions of these smuggler crafts, and then radio the information to the cutters on patrol. This is being done every day, thus preventing the entry of illicit material into the United States. But the most important duty of Coast Guard aviation, from the humanitarian point of view, is the service it affords to the sick and injured far at sea and the scouting for small boats driven from shore.

On August 29, 1916, the president approved an Act of Congress, which authorized the Secretary of the Treasury to establish, equip, and maintain not more than ten air stations along the coasts of the United States at such points as he deemed advisable, and to detail for aviation duty officers and men of the Coast Guard. This act provided that these stations were to be employed in saving lives and safeguarding property along the coasts and at sea and to assist in the national defense.

Selecting a group of young officers, the service sent them at once to the Naval Air Station at Pensacola, Florida, for training, but before these officers could finish their course, the United States entered the First World War. The fledgling airmen were immediately absorbed into the Naval Air Service, performing duty in this country and abroad.

It was not until 1920 that the first Coast Guard air station was established at Morehead City, North Carolina, with planes obtained from the Navy. Despite the lack of facilities and funds, the Coast Guard kept this station in commission until 1921, demonstrating the great value of aviation in the carrying out of Coast Guard duties. Aviation extended the range of Coast Guard humanitarian efforts far out at sea and proved its worth where speed, with its consequent saving of time, is a vital factor.

Coast Guard aviation has continued to grow. Today, the service has air stations situated at Salem, Massachusetts; New York, New York; Charleston, South Carolina; Miami, Florida; St. Petersburg, Florida; Biloxi, Mississippi; San Diego, California; and Port Angeles, Washington. These stations are equipped with long-range amphibian planes constructed in a manner that enables them to land on the roughest seas. Besides these stations, the Coast Guard maintains Air Patrol Detachments at Cape May, New Jersey, and El Paso, Texas. Seven cutters of the Coast Guard fleet are equipped to carry aircraft, which greatly increases the effective cruising range of these vessels.

Primarily, as with the rest of the Coast Guard, service aviation is concerned with the saving of life at sea, and many persons would not be alive today had it not been for the courage of Coast Guard flyers and the sturdiness of their planes. But the saving of life is only one duty with which Coast Guard aviation is concerned. The service as a whole is charged with law enforcement along the coasts and at sea, its aviation playing a vital part in spotting smuggling craft and other vessels which seek to evade the revenue laws of this country.

There are many fishing smacks plying the waters along the northern coasts of the United States, and the work that these fishermen do is hard, cold, disagreeable, and often extremely dangerous. A fall on an icy deck, a slip with a knife, and a man becomes gravely injured. He must be landed for hospitalization at once. But how? The fishing vessel is 150 miles at sea, with rough weather and storm between it and shore. It will take days for the boat to reach port, and by that time, the man will have died. No time to lose. Send out a call for the Coast Guard, ask for a plane to come out, and get the injured man.

Hardly a day passes that the Coast Guard does not receive such a request. And these appeals are heeded, no matter the weather, regardless

of the condition of the sea upon which the plane must land. It is the
Coast Guard's job; they must go out.

Such a call for assistance came from the trawler *White Cap*, on the
stormy night of May 20, 1937, to the commander of the Boston Division.
A member of the *White Cap's* crew had suffered a serious injury and was in
need of immediate hospitalization if his life was to be saved. Orders went
out at once from the division commander to the Salem Air Station, and
at midnight, Lieutenant Commander F. A. Leamy took off to contact the
trawler, which was in a position sixty miles southeast of the air station.

Misty rain made visibility poor; the wind was gusty and rough. Commander Leamy requested the trawler to transmit radio signals, and by
means of the radio direction finder aboard the plane, he flew along the
null of these signals to the trawler's position. A message came from the
White Cap—the patient's arm had been completely severed; he was suffering from loss of blood; all possible speed must be used in getting him
to a hospital.

At 12:40 A.M., the Coast Guard plane was circling over the *White
Cap*, and Commander Leamy noticed that the trawler was laboring considerably in rough west-southwest seas. Making several approaches close
to the water, he realized at once that the sea conditions were extremely
bad and that he might crack up upon attempting to land. Coast Guard
lives were placed in the balance against the life of an unknown fisherman.
If the plane were cracked up, those Coast Guard lives would be forfeited;
if no attempt to land were made, the fisherman would die before the
trawler reached port.

In a few seconds, the commander made his decision. He ordered
flares to be dropped; then, he circled and came in for a landing in the
patch of light that the flares made. The plane nosed closer to the water,
the waves slapping up at the Coast Guard amphibian as it settled down.
Easy, easy, sailor! A false move here, and it's all over. The waves were just
underneath now, huge green things with white lace across their tops.

Commander Leamy eased the plane over the top of one wave and
then brought it down into the trough between the seas. It pounded heavily for a moment, with one wing digging deep into the water, then slowly
righted itself.

The Coast Guardsmen wiped their faces and grinned to themselves. They were down; half of their job was over. They dared not think of trying to rise from that water; it was too easy to see how close to the impossible the task was.

Commander Leamy ordered a quick examination of the plane, but this inspection revealed no apparent damage to the control surfaces and the engine mounts. Once again, Coast Guard equipment had proved its sturdiness. A plane less sturdily built would have crumpled like cardboard under the impact of the seas.

A boat from the trawler had been lowered, and the patient was being brought over. The boat came up under the side of the plane, but the first attempt to place the injured man aboard met with failure owing to the rough seas, which tossed the small craft against the plane and threatened to dump the crew into the water. The Coast Guardsmen put out fenders and, by careful work, managed to get the helpless man into the plane, where he was strapped into a seat and made as comfortable as possible. Making a flashlight inspection of the injured man, Commander Leamy saw at once that the arm had been completely severed just above the elbow. A tourniquet had been placed around the stump, but the patient, already half unconscious from shock and loss of blood, was still bleeding slightly. Realizing that additional first aid would do no good, Leamy made arrangements for an immediate takeoff.

He directed the boat crew from the trawler to keep their boat in the water until the takeoff had been accomplished. Then, looking back at his men, he smiled and opened the throttle.

The plane rode sluggishly through the water, waves beating against it with a sound like that of sails flapping in a gale. It plowed nose first in a breaker, trembled, hesitated, then recovered. It rode over the top of a wave higher than the rest, freed itself momentarily—enough to maintain flight before the next oncoming wave could impede its progress—gained a little altitude, split the tops of the next few breakers, and finally rose from the sea. The men in the boat cheered and waved their hats as the Coast Guard plane headed for Salem Air Station.

Riding the null of the homing beam, Leamy informed the air station of the time of arrival. Back at the station, preparations had already been made in the hangar office for an emergency operation should one be necessary, and an ambulance was standing by outside.

At 2:05 A.M., the plane landed. The patient was examined by a public health doctor, placed in the ambulance, and rushed to the Marine Hospital in Chelsea, Massachusetts.

The courage of Commander Leamy and his crew on that night flight out over the storm-whipped water has been duplicated in dozens of similar cases throughout the years. Commander Leamy and his men came back. Others have gone out but have not come back.

On the morning of April 6, 1939, a Coast Guard plane—carrying Lieutenant Robert L. Grantham as pilot, Radioman James Dinan, Aviation Machinist's Mate Clifford J. Hudder, and R. A. Paddon, another Coast Guardsman—left El Paso, Texas, at about 9:30 A.M., for Galveston by way of Del Rio.

In the afternoon, near Alpine, Texas, they ran into a snowstorm. The plane had been flying over mountainous country, with bearings being taken from a railroad below, when the storm rendered visibility zero. Air currents from the mountains below tossed the plane around like a feather. The air was bitterly cold, and the snowstorm steadily became worse. This was like flying inside of a soap sud, with pieces breaking off and floating past the wildly tossing plane. Lieutenant Grantham was experiencing difficulty in keeping it righted when Hudder, who was seated beside him, pointed out the window at the wing struts. "Ice!" he said. "She's beginning to ice up, sir!" Lieutenant Grantham nodded grimly. Already, the plane was showing the effects of the ice. It was becoming increasingly hard to handle, with one wing persisting in going down. Finally, he turned to Hudder and said: "All right, Hudder—abandon ship."

Hudder stared at him. "But you, sir?"

"Get out," Grantham repeated. "I'll hold the plane until you fellows are clear." His mouth twitched as he fought to get the wing up; he looked at Hudder. "Go along, sailor. See you down below."

Hudder nodded and turned to the other Coast Guardsmen. "Lieutenant Grantham says for us to jump. He can't hold the plane up much longer."

Paddon stared at him.

"Get out!" Hudder ordered again. "Get out!"

With a quick glance at Lieutenant Grantham, Paddon stepped to the door, but the weight of the wind was holding it shut. Finally, they got it open, and Paddon, the first man to go, said softly, "Good luck, fellows!" and let his chute pull him off into the night.

Dinan was the next to jump, followed closely by Hudder. Before he left the ship, Hudder looked back at Lieutenant Grantham and waved. "I'll see you down below, Lieutenant!" And Grantham smiled and called, "Right!"

After Hudder had gone, Lieutenant Grantham felt suddenly lonely. A coldness, not at all from the weather, settled over him. He knew that the plane could not be held up much longer; the weight of the ice was pulling the wing down, down, until the unseen, terrible whirlpool of death was near.

He worked desperately. He got up out of his seat, looked around for something to lash the controls down with, then shook his head quickly. He didn't have time. He would have to get to the door and take his chances on a quick leap into the night. If the plane nosed over and missed him, all would be well; if it came his way—well, that was the chance that he must take.

Clenching his teeth, he whirled and made a dash for the door. The plane lurched suddenly, throwing him against the door-facing. He scrambled up desperately, lost his footing, and almost went down again. By this time the plane was turning over and over, but somehow Grantham forced himself out of the door. For a moment, he thought himself free, then something dark hovered over him.

His half-opened chute was fouled in the falling plane.

Hope vanished.

There was nothing left of life but a howling, twisting fall through the heavens, down through the snow that had caused all this, and finally, the black, red-shot awfulness of the crash. Then, oblivion.

Yes, some come back. Others do not.

Paddon struck on a hill and fell on his back with a force that knocked the wind out of him. He was dragged along for a hundred feet before he could recover himself sufficiently to collapse his chute. This done, he lay down in the snow and tried to rest, though he was wracked with pain. Finally, he managed to get to his feet and started over the hill where he had seen another chute come down.

Hudder came down on another hill not far from where the plane struck. Dinan fell some distance away, and when the three Coast Guardsmen got together, they looked at each other and, without speaking, started for the plane.

They found it in a valley a few hundred yards away. Found it half buried in the snow, crushed and broken to bits. They found Lieutenant Grantham's body wrapped in his parachute, one hand thrown up over his face. The three Coast Guardsmen turned and walked away from the plane without a word; they kept walking until they were a hundred yards away. Then, "I don't believe it," Hudder said quietly. "He can't be dead. I was talking to him just a few minutes ago. He said, 'I'll see you down below.' But he won't. He—"

"We got to get out of here," Paddon broke in quickly. "We got to find help. Night's coming on; the storm's getting worse. We'll get the compass—"

"We won't," Dinan said. "It's broken. I looked at it when we were back there."

The three of them stood silent for a moment while a forty-seven-mile gale howled and whirled the snow around them. They were bitterly cold, and all bore injuries received when they jumped from the plane. Paddon, with a badly bruised back and a twisted knee, was the most severely injured. Every step he took gave him excruciating pain, but he kept to his feet. Once he let down in the cold . . .

The three men drove themselves through the snowdrifts, all sense of direction gone, wandering forward, knowing that they must keep going, keep fighting.

Around five o'clock, they heard the mournful sound of a train whistle in the distance. Hope flared wildly in them, and they started mushing desperately in the sound's direction. For an hour, they kept on, but they did not find the railroad track. Then, as though in mockery, the train whistled again; the sound was weaker now; it seemed to come from behind them. Blankly, they stared at each other.

"We're chumps," Dinan said. "We've been following an echo all this time."

They were completely lost. It was growing dark, and the storm seemed to be increasing. They moved on a few hundred yards, wandering aimlessly, bending against the wind until they could go no further. Suddenly, Hudder discovered that Dinan was not with them. Somewhere, back on the thin trail they had been following, Dinan had turned off, possibly intending to circle around and meet them farther on.

"Let's go look for him," Paddon said wearily.

Hudder shook his head. "He's in better shape than you, Paddon. You can't travel tonight. We'll build fires and use our chutes as windbreakers. Dinan'll see the flames and rejoin us."

So they spent the miserable night in the snow banks, with the chutes spread open and propped up with sticks. They lay down in front of the fire, catching fitful naps while coyotes howled dismally on the hills.

Early the next morning, they started out again, moving forward in a white world, their eyes burning and blurring from the whiteness. Once, during the morning, they thought they heard a plane's motors, but they could not be sure. Dinan failed to rejoin them, and now they feared he must have perished.

In the afternoon, their strength ebbing fast from lack of food, Hudder and Paddon came across an old cattle-dip pit dug in the rugged mountainside. The two weary men dropped down and rested and rubbed snow over their burning eyes.

A noise sounded in the stubby undergrowth nearby. Both Coast Guardsmen whirled quickly. There, standing at the edge of the dip pit, a cow stared at them with lonesome eyes. And behind the cow stood Dinan, grinning from ear to ear.

Hudder said, "Hiya, Dinan. Where'd you pick up the steak?"

"Back up the trail a piece," Dinan said, nodding toward the way he had come. "It's a shame, too. Mooley liked me, fellows, first off."

Paddon grinned. "You guys thinking like I am?"

Dinan nodded again. "Sure. I said it's a shame, but it's got to be done. I found her. Hudder, you got to do the dirty work."

Hudder got to his feet and looked around on the ground until he found an iron rod. He jerked his hand toward the pit. "Let's get her in there. It'll be easier on me and her, too."

Carefully, they herded the cow into the dip pit. Then, Hudder stood astride of the pit and beat down on the cow's head with the iron bar. It was a hard job, for Hudder's strength was almost gone. Finally, however, the cow dropped to her knees, and Dinan cut her throat with a pocket knife.

It took a long hour to skin the carcass and cut off five or six pieces of meat. They cooked the steaks by holding them over the fire on sticks. Paddon said later that, to him, nothing had ever tasted as good as those steaks. They cut off about ten pounds of steak for future use and, having finished their meal, started out again.

That night, the second one after the disaster, they spent in a deserted ranch house. All the next morning, they rested, regaining their strength. In the afternoon, they heard planes roaring in the distance, but there were no visible signs of them. Apparently, the planes were searching on the other side of the mountains, for they did not cross the range. In an attempt to attract them, Hudder and his companions built large fires out in the ranch yard but without success.

The next day, giving up hope of being found, the Coast Guardsmen decided to keep moving. Their trek through the snowbound wastes had partially blinded them; they saw mirages. Dark patches on the snow became ranch houses; waving bushes became men moving along toward them. Finally, the three men came to a little valley between the mountains, and here they halted for a rest.

Hudder said, "If it doesn't turn out to be another dark patch, I think that's a house down there."

Dinan and Paddon stared through swollen lids in the direction of Hudder's outstretched finger.

Dinan shook his head. "I can't see a thing but snow. I don't think I'll ever see anything but snow again."

Paddon nodded quickly. "I see it, too, Hudder. It is a house! There's smoke coming out of a chimney! Let's get going, fellows! "

They ran down the hillside, falling, twisting through the snow, laughing. But when they got down into the valley, the house had vanished into the surrounding snow.

"Gone," Dinan said bitterly. "We're chasing mirages again."

Hudder shook his head. "I don't think so this time, Dinan. Our eyes are bad, that's all. There's a ranch house ahead of us; we'll find it if we keep going."

And keep moving they did, Dinan and Hudder supporting Paddon between them, as his back was troubling him greatly. Finally, when it seemed that they could go no farther, a house loomed out of the snow just ahead of them. A man came out on the porch and stared at the three Coast Guardsmen.

"You're them Coast Guard flyers who crashed with their plane," he exclaimed. "There's a searching party in the mountains looking for you now. I heard about it over the radio."

"Never mind that," Paddon said quickly. "We've got to notify someone of our whereabouts. Got a telephone?"

Without a word, the man led them into the house to a telephone on the wall. Hudder called G. J. McGee, chief inspector of the United States Border Patrol at Alpine, Texas, and advised him of their whereabouts, Mr. McGee replying that assistance would be dispatched immediately.

While they were waiting, the ranchman prepared a hot meal for the Coast Guardsmen and did what he could for their injuries.

Later, they were taken to the hospital, where they remained for some time, recuperating from their terrible experience. All three were quick to state that, had it not been for Lieutenant Grantham's unselfishness and courage, they would have died. As it was, Lieutenant Grantham stayed with the plane, giving the three enlisted men their chance for life, and then, and only then, when it was too late, did he leave his command. He was true to the tradition of the sea and died in that tradition—that the skipper of a ship should be the last to leave her.

From a hospital bed, Paddon said: "At this time, I do not feel I can give a good account of the accident which took the life of one of the service's finest officers. I do want to say, however, that there is no tribute too high for Lieutenant Grantham, for he was 'all man' and gave us our chance for life, even at the cost of his own. . . . It all seems like a nightmare to me, and as yet, I can't fully realize that it happened. . . . I want to state again that the service and the public should know that Lieutenant Grantham had what it takes, and may the Lord be kind to his generous soul."

There's not much more one can say for Lieutenant Grantham. His courage and his unfailing adherence to ideals above and beyond the call of duty have spoken for him. To those who knew him and worked with him, his death was a terrible shock, and the service lost an officer whose fine endeavor and high achievement won him a high place on the roll of the nation's heroes.

Around noon on January 1, 1933, a message was received from Chester Shoals Surf Station by the Coast Guard air station at Miami, Florida— a boy had been swept out to sea in an open skiff in the vicinity of Cape Canaveral, Florida, and two fishing boats had been searching for him since 10 P.M. the night before. Coast Guard aid was urgently requested.

Making immediate preparations to aid in the search, Lieutenant Commander C. C. von Paulsen ordered the big amphibian plane *Arcturus* warmed up and brought out to the line. The commander gathered up the weather reports that had come in and scanned them closely. They were anything but encouraging, with prevailing rain and squalls, and at that time, the wind was blowing from the northwest at a rate of 25 mph.

At 12:20 P.M., the *Arcturus* took to the air, its crew consisting of Lieutenant Commander von Paulsen, pilot; Lieutenant W. L. Foley, co-pilot; Aviation Chief Machinist Mate James B. Orndorff, Jr., mechanic; Aviation Machinist's Mate C William D. Pinkston, mechanic; and Radarman Petty Officer Third Class Thomas S. McKenzie, radioman.

The weather was heavy and gray, with rain being driven down upon them by the wind. Commander von Paulsen headed the *Arcturus* out to sea, then northward toward Cape Canaveral. In about twenty minutes, they arrived at the approximate position of the skiff as given by the Chester Shoals Coast Guard Station. Von Paulsen brought the craft down slightly and flew in ever-widening circles until, at last, the skiff was sighted wallowing in the waves. They saw the boy, too, who stood up in the tiny craft and waved desperately as the Coast Guard plane zoomed over him.

Commander von Paulsen circled the skiff closely, making a careful observation of the boy's condition. The Coast Guardsmen realized at once that the boy was in very poor shape and that his craft was in imminent danger of sinking. It was imperative that a rescue be affected at once.

Swinging out widely, von Paulsen searched for a boat that might be led to the spot, but none was sighted. The nearest Coast Guard craft was at Palm Beach, Florida, some eighty-five miles away. The rescue attempt was clearly up to the *Arcturus*. If the boy lived, it would be through the efforts of the crew alone.

Flying back over the skiff, von Paulsen studied the situation. The wind had shifted to the northeast a few minutes previously and was increasingly squally. It became readily apparent that if the youth were not picked up before nightfall, it would not be possible to locate him again until two or three hours after daylight the next morning. It was definitely established that no vessels were near enough to reach the boy before dusk. Considering these facts—together with the youth's poor condition and the state of the weather—the only possible conclusion to be drawn was that if the boy's life were to be saved, he must be picked up by the plane.

Commander von Paulsen realized that the condition of wind and sea would make the task extremely dangerous and difficult to accomplish. After picking up the boy, the plane might not be able to take off again. He believed, however, that in the event of such a possibility—the *Arcturus* would be able to taxi to shore as it had been built precisely for such type of work. Should the plane become damaged in landing—and thus be unable to taxi—von Paulsen believed that it would stay afloat until driven into the surf.

The question of the safety of the personnel did not enter into the case. This sort of thing was their work; it was their job to die, if necessary, to save another's life. Von Paulsen, veteran that he was in Coast Guard aviation, quickly decided to land and pick the boy up.

After dumping the surplus gasoline, he brought the *Arcturus* in for a landing. Gusts of wind kept whipping the plane up and down, and it was a hard task to keep her on an even keel. With superb skill, von Paulsen set her down in the sea; but a float broke in landing and kept banging against the wing with such force as to threaten grave damage.

Radioman Thomas S. McKenzie, despite the fact that he was aware of the presence of sharks in the vicinity, immediately went overboard to free the float, but fortunately, the wires holding it carried it away, relieving him of the task. Realizing the boy in the skiff was in such a weakened condition that he would not be able to assist himself, McKenzie now swam to the skiff and steadied it while the plane taxied around. Then, by dint of careful maneuvering and with McKenzie's assistance, the skiff was brought under the wing of the plane, and the boy was transferred to the Coast Guard craft.

The absence of the wing float made it advisable to get off the water, if possible. Obviously, if the wing kept dropping down into the sea, the *Arcturus* would soon suffer heavy damage. Commander von Paulsen opened the throttle. The plane bumped heavily through the water, steadied, and took off from the sea like a great wounded seagull. They were in the air!

Misfortune struck swiftly. The plywood on the left wing began ripping away, and the plane could not be held level. Quickly, von Paulsen set the plane down, and this time, the force of the landing wrinkled the hull considerably under the forward spar. There was only one thing left for him to do.

He headed the ship toward shore and attempted to taxi to the beach. However, due to the absence of the wing float—which made it impossible to keep the plane on a desirable course—it was apparent that the try would fail. Realizing this, von Paulsen stopped the engines and ordered a sea anchor thrown out to hold the plane steady. Already the force of the pounding waves was doing its work; the wings were rapidly disintegrating; water was beginning to seep through the plane's hull.

At this time, the sea anchor carried away, causing the plane to lurch heavily, and not until another sea anchor was improvised and put over the side did the craft ride more easily. Directing McKenzie, the radioman, to rig up the radio antenna, von Paulsen now prepared the ship as well as possible for any unpleasant eventualities.

The drift toward the distant surf continued, with the plane's crew making occasional adjustments to the sea anchor to ease the riding. By this time, the plane was showing the effects of the pounding to which she was being subjected. One wing was a mass of wreckage, the other badly damaged. Around 9 P.M., von Paulsen ordered McKenzie to send out an SOS, as no reports of rescue vessels had been received.

SOS! SOS! SOS!

The call whipped through the night, silencing powerful shore stations up and down the coast, bringing countless operators to alert tenseness. The commercial radio stations rebroadcast the message:

COAST GUARD PLANE DOWN OFF BIG BETHEL BEACH
NEEDS IMMEDIATE ASSISTANCE

Coast Guard cutters were dispatched at once to the assistance of the *Arcturus*, but they failed to arrive in time to be of service. At 11 P.M., the seas had increased to a height of fifteen feet with a tendency to break rapidly, and the pounding of the surf was plainly audible to the men in the plane. The water shoaled rapidly, and the craft was pounding. Within an hour, they had passed the first line of surf and were being driven into the second, the sea anchor having failed to hold against the twisting seas.

Commander von Paulsen attempted to start the engines to aid the craft through the maelstrom, but they refused to fire. Those were anxious moments. Finally, at 1 A.M., on January 2, the *Arcturus* successfully passed the last line of surf and was carried high up on the beach, just inside Bethel Shoals—a remote locality entirely surrounded by a primeval jungle.

Realizing that prompt medical attention for the boy was imperative, von Paulsen was about to have him carried on an improvised stretcher through the woods to civilization when a group of Customs Border patrolmen came upon them. Von Paulsen, therefore, sent the boy, in the company of Lieutenant Foley and the patrolmen, to the nearest doctor, where he was placed in the latter's care. Foley then called the office of the Coast Guard East Coast Patrol Area and arranged for the salvage of the plane.

Once again, the Coast Guard had gambled with the sea and won. The stakes—a human life!

Chapter IX

GUARDIANS OF THE SEA LANES

The Coast Guard gambles.

One arm of the Coast Guard, six months of every year, gambles with the sea—with the odds tremendously against it. The International Ice Patrol performs the most hazardous of all feats at sea—its job is to ride herd on icebergs, the most dangerous and treacherous travelers in the sea lanes. Ships flee icebergs; the International Ice Patrol seeks them.

Time and the growing demands of a growing nation have added many duties to the enforcement of revenue laws and cooperation with the Navy in a time of war, until today, the Coast Guard is the country's chief enforcement agency for all maritime laws. The Coast Guard is charged, under law enforcement, with: enforcement of the Customs Laws, prevention of smuggling, prevention of evasion of revenues; navigation and other laws governing merchant vessels and motor boats; harbor rules and regulations governing anchorage of vessels; laws relative to oil pollution; laws relating to immigration, quarantine, and neutrality; protection of game and the seal and otter hunting grounds in Alaska; regulations for the protection of the salmon and other fisheries in Alaska; International Conventions relative to fisheries on the high seas; sponge fishing laws; protection of bird reservations established by executive order; laws generally concerned with Alaska; miscellaneous laws for other branches of the government; and suppression of mutinies on merchant vessels at sea.

Life-saving and assistance became a duty of the Coast Guard in 1831 when the Secretary of the Treasury observed in a written report, ". . . it is thought proper to combine with the ordinary duties of the cutters that of assisting vessels in distress at sea, and of the administering to the wants of their crews." Since that time, this duty has grown to be the most dramatic function of the service, and daily, in the nation's newspapers, may be found accounts of the Coast Guard's efficiency in life-saving; daily, literally hundreds of persons are rendered assistance, ranging from the transportation of cattle from flooded lowlands to the removal of a critically ill seaman from a freighter at sea.

Along the coastlines of the United States are Coast Guard surf stations, manned and ready for any emergency that might occur on the sea. Watches are maintained day and night, and alert eyes scour the sea for any sign of distress. A canoe may be overturned, with the occupants waving frantically from the water. A lifeboat goes out immediately; the victims are picked up and more lives are added to the growing roster of those saved by the Coast Guard. Or, perhaps, the emergency may come in the form of a winding thread of smoke, a sudden gush of black against the blue of the sky. Flames in the night. People jumping from the deck of a burning liner, without a thought except to escape the consuming tongue of Hell. Explosions . . . screams. Burning oil on the sea. And through it all, Coast Guard lifeboats plow, picking up and landing survivors and going back time and time again, searching until there are no more cries for help, only the sodden slap of the waves against the sides of the surfboats.

All this comes under the insurance that the Coast Guard gives to the people who travel the sea. Cutters and patrol boats are stationed at strategic points along the coasts, with radio receivers manned by specially trained men waiting for the call that may be a simple request for weather reports or a frantic call for help. Cutters are equipped for towing so that in case a ship becomes disabled, it may be towed in to the mainland within reach of commercial tugs.

One of the greatest humanitarian duties assigned to the Coast Guard is the International Ice Patrol, which came under the jurisdiction of the service in 1913, just a year after the *Titanic* disaster. Since that time, with the exception of the war years—1917–1918—the Coast Guard has maintained an annual patrol of more than 40,000 square miles of sea, an expanse equal to the area of Pennsylvania.

On the night of April 14, 1912, the palatial British liner *Titanic*—on her maiden voyage from England to the United States, with a passenger list of 2,223 men, women, and children—was just south of the Tail of the Grand Banks of Newfoundland, speeding toward New York at more than twenty knots. The night was perfect, with visibility through the crisp, clear air practically unlimited.

The great engines were throbbing, singing praises to man's conquest of the sea. With her steel hull and her watertight compartments, the ingenuity of man had produced a craft that would withstand the beat of the elements at their worst, ensuring safety for the thousands of persons who would take passage for the quick, racing trips across the Atlantic.

But Fate was shadowing that lean greyhound of the seas that night. For even as the gay passengers laughed and talked and danced, a low-lying iceberg, just visible over the top of the water, came drifting out of the north to lie in wait for the dashing *Titanic*.

The liner *Carpathia*, some fifty or sixty miles away, had been in radio communication with the *Titanic*, and the grim words had gone forth:

ICEBERGS IN YOUR IMMEDIATE VICINITY.

But the speed of the *Titanic* was not diminished. There was the speed record to think of, a record that would entice many more passengers to span the Atlantic aboard the great *Titanic*.

Then, suddenly, as the dance bands played and merriment and good fellowship reigned, the iceberg hove up abruptly in the path of the speeding liner.

Immediately, the cry came from the lookouts: "ICEBERG DEAD AHEAD!"

Orders were snapped on the bridge. Even then, no particular anxiety was felt by the officers, as the iceberg had been sighted in plenty of time. The wheel went over, full astern was signaled down to the engine room, and the *Titanic* gracefully answered these commands.

The officers looked at each other and smiled. How well the great ship answered her helm! As light as a feather to the touch, she was!

Then—disaster!

The hull of the *Titanic* crashed upon the portion of the iceberg hidden beneath the surface of the water, a portion five times as large as that which was above water. The steel crumpled like paper from the force of the collision, ripping and tearing the hull completely away, opening great wounds through which the sea poured in torrents. At that moment, the great *Titanic's* engines stopped throbbing; she became a sluggish, ungraceful thing, broken by her joust with the sea.

Mere words can never describe the horror of that night. Great deeds were done, deeds of unstinting sacrifice and heroism. Men stood back and watched their wives and children climb into lifeboats, knowing full well that there would not be enough boats left to carry them away from the doomed liner.

The *Carpathia* heard the *Titanic's* distress call and immediately started full speed for the distress position. The word went around on the *Titanic*: "There's another boat coming. They'll be here in a short while, and we'll all be saved!"

Her lifeboats, seventeen in number, had cleared the side of the sinking ship, carrying mostly women and children. But hundreds remained aboard, grouped about on the decks, while the bands played on and on to keep cheer and hope alive. And while anxious eyes strained toward the horizon for the first sight of the rescue boat, prayers were offered up.

Within an hour after the collision, the *Titanic* was well down by the bow, and as the water ripped through the watertight compartments with gurgling sounds, explosions rocked the ship. The officers realized now that she would never stay afloat until the *Carpathia* arrived.

Already, the decks were slanting and water sloshed not far below the rails. A grim resignation settled over the people on the *Titanic's* decks. Husbands and wives stood holding hands, looking at each other, thinking of the past, wondering as to the future, in the so little time left to them.

Then, three hours and forty-five minutes after the collision, the *Titanic* reared up and, with a series of explosions, slid beneath the water—1,517 people perished in this disaster, one of the greatest in all maritime history.

The world was horrified. The maritime nations arose as one and demanded protection for their ships and the passengers who traveled in

them. Wild schemes for the destruction of the icebergs were solemnly discussed, schemes which eventually fell through for their lack of feasibility.

The United States government saw one way of combating the ice
monsters, however, and within a month after the sinking of the *Titanic*,
two cruisers—the *Chester* and the *Birmingham*—were assigned to the
ice-infested area. These vessels remained on patrol duty throughout the
ice season, until the bergs were no longer drifting down into the steamship lanes.

The next year, in 1913, the United States Coast Guard took over this
duty, and it is a matter of national pride that since that time, not a single
life has been lost as the result of a ship colliding with an iceberg.

The International Ice Patrol is the direct outcome of the International
Conference for the Safety of Life at Sea, which was held in London in the
fall of 1913, and which resulted in a treaty being drawn up among the
thirteen maritime nations vitally interested in trans-Atlantic shipping.
This treaty was signed in January 1914 and provided for an international
derelict destruction, ice patrol service, and observation, with the United
States being invited to assume the management and operation of this
service. The expenses of the Ice Patrol were to be defrayed by the several
governments involved in a proportion determined solely by their ocean
tonnage. The United States, at this time, contributes about 18 percent of
the expenses of the Ice Patrol.

The Ice Patrol cutters depart for the ice fields just before the berg
season starts, which is usually during March, and maintain a continuous
patrol until the latter part of June or early in July when a final survey of
the ice area is made to make sure that no more bergs will reach the steamship lanes. These cutters, of the large, long-range cruising class, alternate
in patrolling, each taking a fifteen-day tour of duty. Upon being relieved
by the other, the cutter proceeds to Halifax, Nova Scotia, the patrol's
base, for fuel and supplies.

Since the ice area is manifestly too large for the cutter on duty to
cover within a useful time, it has been arranged with all vessels passing
through the area to send in radio reports to the cutter every four hours,
giving the location of any iceberg sighted, its size, drift, and the temperature of the water at the surface in its vicinity. All this information

is used by the ice observer officer in making up current and drift charts, which tell at a glance the location for every minute of the day and night of any iceberg in the patrol area, together with its direction of movement and drift.

Gathered from the four corners of the ice area, this information is broadcast to shipping four times a day and is available between broadcasts to any ship requesting it. The same information is transmitted to the Hydrographic Office of the Navy Department, in Washington, D.C., for dissemination in the daily bulletins issued by that office.

The radio room aboard an ice patrol cutter becomes, therefore, a general clearing house through which all information relative to ice conditions passes. Six radio operators, specially selected for ice patrol duty, alternate in four-hour watches during the day and night, handling an average of 90,000 words transmitted and received during the fifteen-day patrol. It is a nerve-wracking job, for one mistake in the transmission or reception of ice reports may send some great passenger liner, carrying thousands of persons, crashing against the side of an iceberg.

In order that all data relating to ice conditions may be properly assembled and prepared for release on the daily broadcasts, an officer having a special aptitude for this duty and who has made an extensive study at one of the nation's leading universities on the subject of oceanography, the icebergs, their properties, formation, drift, and disintegration, is detailed by Coast Guard Headquarters as the ice observation officer. This officer goes out on the first patrol and thereafter transfers to the relieving cutter, taking with him all data and accumulated reports so that he has a continuous personal knowledge of all phases of the Ice Patrol ready and at hand at any time during the entire ice season.

Nearly all the icebergs which appear near the trans-Atlantic steamship lanes south of the Tail of the Grand Banks come from the west coast of Greenland, between Disco Bay and Melville Bay, breaking away in the summer and autumn to travel 1,800 miles to the latitude of the Banks by the following spring.

With the exception of its southern area, Greenland is covered with an enormous sheet of ice, some five to seven thousand feet thick; its valleys are filled with glaciers that move slowly westward to the sea.

During the summer months, some of these glaciers move forward at the rate of sixty to seventy feet a day, and as they slide out into the sea, the weight becomes too great for them to remain intact with the rest of the glacial mass, and they break off with resounding crashes, throwing spray and ice skyward in a series of explosions.

Thus, icebergs are born. Of the thousands of bergs to start on the long journey southward, only a small proportion finally drifts into the Atlantic to become the plague of shipping on the northern steamship lanes.

In some seasons, the icebergs reach the shipping lanes in greater numbers. In 1922, approximately 1,200 bergs drifted south into the patrolled area, while in 1924, only eleven appeared. This great disparity may be traced directly to weather conditions in Iceland, and yearly, the experts on the Coast Guard cutters are able to predict to an astonishing degree of accuracy the number of bergs that will drift down into the lanes. If the winter and spring in Greenland have been unusually severe, with cold weather lasting far up into the summer, the flow of ice down through the valleys to the sea is slowed, thus resulting in a decrease in the number of bergs broken off in the water.

During a normal season, however, from 300 to 350 icebergs come south from their place of birth, but of this number, only forty or fifty manage to survive the long trek down as far as the Tail of the Grand Banks. In fact, only one berg in every four years gets as far south as the fortieth parallel of latitude. Several instances have been recorded where Arctic ice has been sighted as far south as the Azores and Bermuda, but such cases are rare indeed.

After the middle of June, there is a steady decrease in the number of bergs in the ice area, and from the middle of July until the following March, the ocean is practically devoid of ice. After a survey around the coast of Greenland to determine if any more bergs are likely to break away in sufficient size to survive the trip down into the lanes, the Ice Patrol is officially over, and the cutters get underway for their home ports.

Through years of experience in the ice fields, with all relative data carefully compiled and studied, Coast Guard Ice Patrol officers have determined that, while icebergs are naturally affected by the shifting winds across Baffin Bay, it is the various ocean currents that govern to a marked degree the direction of drift south from Greenland.

The main current which brings these huge masses of glacial ice down upon the Banks is known as the Labrador current, coming south from the icy waters of the Arctic Ocean and flowing through Baffin Bay and Davis Straits to the west of Greenland, whence it follows the coasts of Labrador and Newfoundland. In the vicinity of the Banks this current branches, the main portion curving to the eastward and following the eastern edge of the Grand Banks. At the southeastern corner, another division occurs, with the minor portion curving toward the west and the main body continuing south until it is re-curved upon coming in conflict with that great ocean current known as the Gulf Stream.

As the general drift and the rate of speed of these ocean currents control the movements of the icebergs, current charts are constantly being compiled by the ice observer on the patrolling cutter, both from the data recorded on board and from information received from other ships passing through the ice area and regularly reported by radio every four hours. From this assembled information, it is possible to estimate the daily flow of the ice wall along the eastern branch of the Labrador Current and, in this manner, to determine the approximate drift of the bergs.

For the past several years, the Coast Guard has detailed another cutter, in addition to the two assigned to the Ice Patrol, to make a special ice observation cruise along the west coast of Greenland, and many bergs have been sighted, both at and near their place of origin. These bergs were from 250 to 500 feet high, with possibly 2,000 or more feet under the water. The largest iceberg sighted by these expeditions was nearly 1,700 feet long but only 60 feet high and was estimated to contain nearly 40,000,000 tons of glacial ice.

In icebergs, nature sometimes turns loose a frenzy of architectural ingenuity and, with the force of the wind, the beat of the sea, and the heat of the sun, accomplishes some weird and beautiful sights. One Ice Patrol cutter came upon an iceberg moving serenely along on the broad Atlantic that looked for the world like the British Lion. In the bright sunlight, these bergs make an inspiring sight, with their glazed whiteness catching and reflecting the sun's rays like a huge jewel. However, all bergs are not completely white. Some of them are streaked with tints of the most delicate blue, which is caused by the strains of purest glacial ice. Other bergs appear to be gowned in capes of studded diamonds as the melting ice falls down through the sunlight to the sea.

A common type of iceberg sighted in the patrolled area is the "Dry-dock," which gains its name from its marked resemblance to a floating drydock and has two sides rising high in the air, each side usually joined with the other beneath the water. Another type of iceberg is known as the "Growler"—a remnant or a broken fragment of a berg.

From the crow's nest of a ship on the rare days in the North Atlantic when visibility is good, an iceberg may be seen at a distance of ten to fifteen miles and, in exceptional cases, as far as twenty-two miles. The watch on the bridge of the ship should, under the same conditions, pick up the berg while it is eight miles off. On moonlight nights, with the weather clear, icebergs may be seen at a distance of eight to ten miles and are plainly discernible to the naked eye as glistening, luminous objects, exquisitely beautiful under the soft rays of the moon. But the type that sank the *Titanic*, a thin shaft of whiteness lying low in the water, is hard to see at distances of more than a mile and is particularly dangerous to approach because of its knife-like edges which extend out beneath the water over a considerable area.

In foggy weather—prevalent along the Banks most of the time—a berg can be seen for only a short distance; it appears as a blurred shadow, a counterpart of the swirling, gray mists.

Contrary to popular belief, the Coast Guard does not undertake to destroy the icebergs in the steamship lanes. During the first few years after the *Titanic* disaster, the Coast Guard was swamped with suggestions from well-meaning but ill-informed people as to how the ice monsters might be destroyed. These suggestions ranged from the ridiculous to the pseudoscientific. Numerous experiments have been done with dynamite with little success. Termite has been used with moderate success, but the cost of such blasting as compared with the results obtained is prohibitive. The Coast Guard has taken the only logical way of combating these navigational menaces, and that is by riding herd on them, broadcasting their positions continuously from the time they enter the lanes until the warmer water of the south melts them.

Each year, on April 14, the cutter on duty in the ice fields carries out a solemn and inspiring ritual over the spot where the *Titanic* sank with 1,517 souls in the early morning of April 15, 1912.

Approaching the scene of the disaster, a strange mood sweeps over the men of the Ice Patrol Cutter. They gather up their "acey-deucey" boards, that time-honored game of the sea, and put away their books to collect at the rails beneath which the water turns up in foaming whiteness.

Quartermasters coming down from the bridge are stopped and asked time and again: "Where are we, mate? Near the spot now?"

A bell jingles in the engine room. The great engines slow down and finally stop, and the cutter rolls gracefully in the swells. Recruits, men who have never made the Ice Patrol before, stare at each other. The grim realization that here, at this exact spot, over a quarter of a century ago, a great ship sank, sweeps over them; that here, in the vast stillness of the morning hours, men, women, and children died with nothing but echoes answering their cries for help.

Somewhat subdued, a boatswain's pipe shrills, calling the men to formation. The Coast Guardsmen gather quickly and quietly on the quarterdeck, dress uniforms spotless, shoes glistening. Overhead, the church pennant flutters while they stand motionless, listening to the calm, firm voice of the cutter's commanding officer as he reads the awe-inspiring burial service of the sea.

As he finishes, three rifle volleys roll out over the gray wastes in measured succession. The formation is then dismissed. Another service in memory of the 1,517 persons who died aboard the *Titanic* has been concluded.

The Coast Guardsman's life aboard an Ice Patrol cutter at sea is not as bad as it may seem. The ship carries a movie machine, and movies are held every night, with ice cream and coffee being served after the show. Among the crew of every Coast Guard cutter may be found talented musicians; every night on the Ice Patrol, weather permitting, the Coast

Guardsmen who are off duty gather on the berth deck to hold a community sing.

Food is always plentiful, and good meals and movies go a long way toward breaking the monotony of the dreary days. At the end of the Ice Patrol, the two cutters meet at a rendezvous agreed upon, and a boat race is held between picked crews from each ship. There is heavy betting, the men of each cutter backing their respective crews with money and good-natured argument.

Another duty has been assigned to the Ice Patrol during the past year. With the installation of air surveys in connection with the establishment of passenger plane routes over the North Atlantic to Europe, the Ice Patrol cutters have been especially equipped for meteorological observations along these routes. Daily balloon ascensions are made with tiny radio transmitters, whose signals are recorded below. By the interpretation of these signals, the humidity, temperature, and pressure at any height up to that at which the balloon burst is determined. Accurate recordings have been made at an altitude of nearly fifteen miles.

A specially trained officer, Lieutenant N. W. Sprow, has been assigned to this work, and he and his assistant, Chief Radioman Holden, have, under the most trying conditions, compiled various meteorological data which will prove invaluable to the safety of the giant clipper ships on their flights to and from Europe.

The Coast Guard is justly proud of its Ice Patrol record. More than 2,000 ships of all types and sizes ply the waters of the North Atlantic each spring, passing close to the dangerous ice fields, where giant bergs lurk in wait for the unwary ship. It is the Coast Guard's job to keep a tab on those bergs, never letting them escape from under the death watch placed upon them by the Ice Patrol cutters. It is a weighty responsibility on the Coast Guard's shoulders, but the manner in which the service has upheld that responsibility fills one of the brightest pages of the nation's history.

Chapter X

DOTS AND DASHES

DIT DIT DIT   DAH DAH DAH.   DIT DIT DIT!
The spine-chilling, quick-breaking series of dots and dashes spins out of the ether, snapping in the ears of hundreds of Coast Guard radio operators. With these signals, a well-rehearsed plan of operation is put into immediate effect by the Coast Guard communications system. It is an SOS, the dreaded call of the sea, that takes precedence over everything—a call for which the Coast Guard radio operator is ever on the alert!

From his first days in the radio school at New London, the Coast Guard operator is taught to regard the efficient handling of distress at sea as his primary duty; he must become the eyes and ears of a service whose job it is to save lives and property on the high seas.

The Coast Guard operator's job is highly specialized. He must be able to recognize any signal on the air that might indicate distress, however minor, on a ship far from shore. Throughout the day and night, he sits at his operating desk on board a cutter or at a shore radio station, listening to the signals flying back and forth through the air, waiting for the one that will send a cutter or a plane dashing out to render assistance.

The Coast Guard maintains a continuous watch on five hundred kilocycles—the international distress frequency—the operator on watch being required to make an entry in his log of some signal he has heard within the last three minutes. Sometimes these logs tell the story of death and heartbreak, with incidents of untold suffering and self-sacrifice written between the lines.

Upon the receipt of a distress call on board a Coast Guard cutter, the radio operator presses a button underneath the desk, thus informing the quartermaster of the situation. The latter immediately calls the officer of the deck and the radio operators off watch, who report to the radio room. One operator takes over the Coast Guard calling frequency, handling all traffic coming in from other units and radio stations; a third takes a station at the radio direction finder, ready to get a bearing on the distressed ship's transmitter signals; the fourth stands by to be of any assistance he can in the radio room.

Once the position of the ship in distress is determined, this information is forwarded to the district headquarters for dissemination to other cutters, which may be within a more effective range of the distressed ship. Often, the Coast Guard's efficiency with the direction finder means the difference between success and failure in a rescue case. The operators are trained to get a bearing within a minimum of time, and usually, such bearings are strikingly accurate.

In the case of the Navy dirigible *Akron*, however—which departed from Lakehurst, New Jersey, at 7:30 P.M. on April 3, 1933, for the purpose of calibrating radio direction finder equipment—rescue operations were greatly hampered by static and electrical disturbances. Around 8:45 P.M., when the *Akron* was about thirty miles south of Philadelphia, thunderstorms were sighted and the giant airship was headed east-northeast. She arrived over the Jersey shore at about 10 P.M. with lightning flashing around her. In order to avoid the growing storm, her course was changed to west, and she continued in that direction until about midnight. At this time, a light was sighted on the ground, possibly Absecon, and again the course was changed to 130 degrees. Around 12:30 A.M., the huge craft apparently entered the center of the storm; she began to toss violently. The force of the storm tore the controls away, and with the gallant Navy men trying desperately to save her, she started downward like a crippled bird. Nothing could be done. The descent toward the water continued. Then, suddenly, a terrible impact demolished the cabins, trapping men in the wreckage, holding them while the water closed over them, and it was no longer possible to fight.

Meanwhile, the German tanker *Phoebus*, bound for New York, was about twenty miles east of Absecon Light when she sighted the *Akron's* descending running lights. Changing her course, the tanker headed toward the airship and, at last, came upon the huge hulk in the darkness. Boats were lowered in a search for survivors. An attempt to make the *Akron* fast to the side of the *Phoebus* was unsuccessful, and the crippled airship drifted away into the darkness. Fortunately, however, the lifeboats managed to pick up the *Akron*'s executive officer and three enlisted men, and despite the gale and heavy seas, the search for the other survivors was continued.

Around 1:30 A.M., the *Phoebus* tried unsuccessfully to raise the Tuckerton, New Jersey, radio station. However, the operator, hearing the Montauk radio station, called and raised it. Two minutes later, the message that was to shock the world flashed into the air:

WSE WSE DE DDPF RUSH MESSAGE NAVY DEPARTMENT
BT AIRSHIP AKRON AFLOAT OFF BARNEGAT
LIGHTVESSEL PICKED UP SOME CANT GET THEM ALL
CHIEF OFFICER THREE MEN SAVED BT MASTER

Because of the heavy static existing, the *Phoebus* experienced considerable difficulty in clearing this message to Montauk. The Coast Guard destroyer, *McDougal*, anchored at Sandy Hook, managed to copy the original message through the crashing static, but other more distant Coast Guard units were unable to copy it fully.

Montauk Radio—WSE—upon receipt of the message from the *Phoebus*, immediately forwarded it to the Mackay Office in New York for further transmission to the Navy and Coast Guard. The *McDougal*, however, cleared the information through the Coast Guard radio station, NMY, at Rockaway, New York, and the news of the disaster reached the communications office of the New York division just seven minutes ahead of the message from the Mackay office.

Rockaway Radio, Coast Guard, broadcast the information on Coast Guard frequencies around 2:03 A.M., and by 2:15 A.M., the Coast Guard destroyers *McDougal*, *Hunt*, *Cassin*, and *Tucker*, and the cutter *Mojave*, with several patrol boats, were on the way to assist. The destroyer *Tucker*, bound for New York, was in the general vicinity of the *Akron*'s position; she changed her course, immediately heading at full speed for the scene of the disaster. At 5:15 A.M., the *Tucker* took radio bearings on the *Phoebus*'s signals, and despite the fact that the static made these bearings highly doubtful, the *Tucker*'s commander changed his course in accordance with them. Half an hour later, the *Tucker* sighted a yellow life raft and much floating wreckage—obviously equipment from the *Akron*. The nature of this wreckage indicated that the airship had struck with terrific force, tearing the cabins and understructure to bits. Circling the wreckage slowly, the *Tucker* searched for survivors, but none were found. At 6:05 A.M., the *Phoebus* was sighted, and the *Tucker*, proceeding close alongside her, established communication with Lieutenant Commander H. F. Wiley, the *Akron*'s rescued executive officer. From him, the *Tucker* learned that one of the survivors was critically injured; all needed medical attention.

Despite the heavy seas, the *Tucker* put a boat over the side and sent a pharmacist's mate over to the *Phoebus*. Meanwhile, at about 6:50 A.M., the *Mojave*, the *McDougal*, and the Coast Guard plane *Antares* arrived on the scene, guided by the bearings taken on the *Phoebus*'s radio signals. The *Mojave* lowered a surfboat for the purpose of sending a chief pharmacist's mate over to the *Phoebus*, but before the boat reached the tanker, word was received that the unconscious man, Chief Radioman William Coplin, U.S. Navy, had died.

The *Tucker* took on board the body of the radioman, along with Boatswain's Mate Second Class Richard E. Deal, who was injured, and Moley Erwin, metalsmith. At 7 A.M., Lieutenant Commander Wiley was taken aboard. Then the *Tucker* got underway at 25 knots for the Brooklyn Navy Yard, where she arrived at 12:30 P.M.

While rescue operations in this case were greatly hampered by the prevailing storm, which caused heavy static up and down the coast, the outstanding feature of the *Akron* disaster was the manner in which the Coast Guardsmen handled communications under the most trying circumstances. At New London, Connecticut, the logs at the radio station stated: "Lightning very bad, breaking down across transmitter, noise terrific." It was almost miraculous that the distress message from the *Phoebus* was copied at all. Certainly, it was a credit to the ability of the radiomen, who stand the long, tedious watches aboard the Coast Guard cutters and stations, that such accurate bearings were taken on the tanker's wavering signals.

On January 15, 1932, the Coast Guard destroyer *Herndon*, on patrol approximately fifty miles southwest of Montauk Point, encountered fog at 12:30 P.M. and reduced speed to 15 knots. Fog signals were immediately sounded and were continued until 12:44 P.M. when the *Herndon* was stopped to ascertain the location of a fog signal which had sounded off the port bow.

Suddenly, the fog horn bellowed again, this time close at hand. The officer on watch on the bridge of the destroyer immediately signaled the engine room for full speed astern, and the powerful engines had just begun to throb when a vessel heading right for the Coast Guard destroyer loomed up in the fog.

Whistles shrilled on the deck of the approaching ship; its fog horn blasted frenziedly. Too late! Her great bow crashed into the port side of the *Herndon*, entering just aft of the wardroom pantry, ripping the plating off the pantry and office, and breaking through number one fire room for a distance of about six feet. At the same time, the side of the bridge just above the pantry was smashed in and the radio direction finder dismantled.

The ship, her identity later established as the *Lemuel Burrows*, bound from Boston to Newport News, Virginia, backed away from the *Herndon*, and immediately the Coast Guard destroyer began to fill with water forward of number two fire room. As the *Herndon* had been standing on number one fire room, the water, reaching the fires, soon rendered the destroyer helpless and placed her in grave danger.

The radioman on watch in the radio room had just finished transmitting a message to Rockaway Radio and had jotted the time on the message blank when the *Burrows* crashed into the side of the *Herndon*. The impact threw the operator out of his chair and broke several tubes in the receivers, rendering this equipment inoperative. The radioman tried to start the transmitters, but the bow of the *Burrows* had cut all the cables leading from the engine room and the auxiliary power supply to the radio room.

Immediately after the collision, the chief radioman and the operators off watch entered the radio room. These men went to work at once to get equipment operating, for word had come up from the engine room that the water was rising rapidly and that assistance must be had at once.

The *Burrows*, in the meantime, had lowered a boat and was searching in the fog for the *Herndon*. The former's radio operator kept sending the question: "WHAT SHIP JUST COLLIDED US?" The *Herndon* was, of course, unable to answer since her transmitters were out of commission. However, the Coast Guard radio station at New London, Newfoundland, heard the *Burrows* and immediately placed two men on watch.

Two minutes later, the New London Station of the Radio Marine Corporation, WSA, transmitted the following on 500 kilocycles, the International Distress frequency:

NCU NCU DE WSA BT FOLLOWING RECEIVED FROM
WJCS SS LEMUEL BURROWS QUOTE WHAT SHIP JUST
COLLIDED WITH WJCS UNQUOTE

This was copied by the Coast Guard radio station at New London and by the Coast Guard patrol boat *Marion*, at anchor in Old Harbor, Block Island.

Then, just four minutes after the collision, the *Burrows* came on the air again and broadcast on 500 kilocycles the following:

CQ CQ DE WJCS SS LEMUEL BURROWS JUST RAMMED
COAST GUARD SHIP ABOUT SIXTY MILES SOUTHWEST
BLOCK ISLAND LATITUDE 40-10 N LONGITUDE 72-22 W
THICK FOG

Three Coast Guard units recorded this message in its entirety, and at 12:51 P.M., Rockaway Radio transmitted the silence signal on the Coast Guard calling frequency, indicating that a distress call was out and for all Coast Guard ships to remain quiet and listen.

At 1:02 P.M., the Southampton radio station called the Coast Guard patrol boat *Marion* and asked her if silence was desired on the distress frequency. The *Marion*, busily trying to contact the *Lemuel Burrows*, replied in the affirmative. Accordingly, Southampton Radio, WSL, broadcast the following:

CQ CQ DE WSL WSL CG NRLY WANTS ALL TRAFFIC
SILENCED QRT [stop sending] DE WSL PER NRLY

Eleven Coast Guard units received this message, which served to acquaint them with the situation. Additional men were placed on watch on the Coast Guard vessels. All operators stood by alertly, waiting for any information concerning the distressed Coast Guard ship.

Sometime later, the *Marion* called the *Lemuel Burrows* and asked what her condition was and the name of the Coast Guard vessel the *Burrows* had rammed. The *Burrows'* operator answered that, due to heavy fog, he did not know but that a boat was trying to locate the rammed Coast Guard vessel. At 1:11 P.M., the *Burrows* called the *Marion* and said, obviously in error, that the Coast Guard vessel destroyed was the *Hartland*.

Meanwhile, the radiomen aboard the *Herndon* had been working desperately to get the radio equipment in operating condition. An extension cord was used to connect the gasoline-driven Kohler auxiliary power supply to the transmitters, alternating between the low-frequency and the high-frequency equipment, in order to be able to communicate on both the Coast Guard and distress frequencies.

The Kohler plant on the Herndon had been used as an auxiliary ship's lighting plant in addition to its primary function as an emergency radio supply; its automatic governor had been disconnected, and the machine set for maximum output, which was 120 volts at 90 amperes. Consequently, the load of the transmitters was not enough to hold the generator's output down to 120 volts, and approximately 160 volts were thrown across the radio motor generator.

As the extension cable was not designed to carry such a current, it became hot, and three fires were started in the radio room in the fuel oil, which was sloshing around on the deck. However, this arrangement was used until the cable burned out.

At 1:14 P.M., twenty-nine minutes after the collision, the *Herndon* came on the air and transmitted a distress message on 500 kilocycles:

SOS SOS DE NRDL NRDL CG DESTROYER HERNDON
MONTAUK 25 DEGREES FIFTY-FIVE MILES DISTANT
MONTAUK POINT MY RECEIVER OUT OF
COMMISSION ANSWER 2675 KILOCYCLES.

Ten Coast Guard units received this message, and five minutes later, when it was transmitted on the Coast Guard calling frequency, 2675 kilocycles, nine more copied it. The Coast Guard cutter *Acushnet*, at Woods Hole, Massachusetts, got underway immediately for the *Herndon*'s position.

The *Burrows* called the *Marion* and asked him to inform the *Herndon* that a small boat was out searching for the destroyer. Around 3 P.M., the *Burrows* found the *Herndon* settled low in the water and took her in tow.

The *Burrows* towed the *Herndon* to a spot just inside of Montauk Point, and at 6 P.M., the *Acushnet* contacted the two ships and took over the towing of the *Herndon* into port. No one was fatally injured in the collision, although four members of the *Herndon*'s crew received painful wounds.

This case gives evidence of the initiative that Coast Guard radiomen are sometimes called upon to exert in their daily duties. The fact that the radio equipment on the *Herndon* was totally disabled by the collision and within thirty minutes was operating testifies that the service radioman is trained for any emergency.

On March 9, 1928, the patrol boat *Bonham* was on patrol off the New England coast when orders were received to seek shelter as a heavy storm was heading up the coast. Steering for Gloucester, the Bonham arrived there at 5:06 P.M., and, secured to a dock, awaited the storm. At 7 P.M., a snow-laden northeast gale was blowing.

At 9:30 P.M., the *Bonham* received orders to proceed in company with the *Active* to the assistance of the passenger liner *Robert E. Lee*, which was aground on Mary Ann Rocks. In ten minutes, the *Bonham* was underway, heading out into blinding snow and wind, which was steadily increasing to whole gale force. A heavy sea was running, but the staunch little boat kept battling on until, at 2:50 A.M., she reached the stranded steamer.

After looking the situation over, Boatswain Brown, in charge of the *Bonham*, decided to wait until morning before attempting rescue work. In making this decision, he was guided by the fact that the *Lee* was not in immediate danger, as she was grounded on the lee shore, and any attempt at a rescue would probably be attended by loss of life.

At daybreak, the wind was blowing a whole gale with heavy snow, which blanketed the *Robert E. Lee* from view. Around 8 A.M., however, the weather cleared, and the stranded ship could be seen well inshore, with heavy seas breaking around her. Boatswain Brown saw at once that the use of a breeches buoy was impossible, for the *Robert E. Lee* lay too far offshore. Accordingly, he decided to move in closer to investigate and to report conditions to the *Tuscarora*, the senior ship on patrol.

As the *Bonham* moved slowly toward the *Robert E. Lee*, a message was received from the *Tuscarora* asking if the *Bonham* could get in alongside the stranded vessel. Brown replied that he would try, adding that he would keep the *Tuscarora* advised as to his progress.

The *Robert E. Lee* advised the *Bonham* through the *Tuscarora* that the starboard quarter afforded a slight lee but that there was a large reef just amidships. The *Bonham* sighted another large reef about a hundred feet off her starboard quarter and, using all precautions, tried to swing around so as to head into the heavy seas and gale in getting alongside the *Lee*. This operation being unsuccessful, the starboard anchor was dropped, Brown intending to veer cable. Owing to the rocky bottom, the anchor dragged for a considerable distance, bringing the *Bonham* nearer and nearer to the reefs. Finally, however, the anchor took hold and assisted the engines in swinging the ship around.

The anchor was now hove up. Boatswain Brown maneuvered the *Bonham* carefully down between the reefs as heavy seas broke and foamed across the Coast Guard boat. It was ticklish work; one false move and the *Bonham* would be thrown against the reefs. In that sea, she would not survive much pounding before becoming a total wreck.

Finally, however, *Brown* got the *Bonham* alongside the *Lee*. While he held the patrol boat's bow against the stranded ship's side, a Coast Guard seaman stood on the pitching deck with a line in his hand. At a signal from the *Bonham*'s skipper, the seaman hurled the line with all his strength toward the *Lee*'s deck high above him.

The coils flew out. The line went straight to its mark, fell across the *Lee*'s rail, hung there for a moment, and then it slipped downward as the *Bonham* was forced toward the reef. The seamen on the *Lee* fumbled and missed it, thus allowing the *Bonham* to drift dangerously close to the hidden rocks.

Signaling for the hawser to be released, Brown at once set about the task of extricating his vessel from its precarious position. Full speed astern was signaled to the engine room. While the passengers watched from the *Lee*'s rails, the Coast Guard vessel backed slowly seaward.

Three times the Coast Guard officer risked his vessel in attempting to get close enough to the *Lee* to remove the passengers and crew. Each time, he missed success by a matter of inches, finally being compelled to abandon the attempts, as the *Bonham*'s engines lacked the power to buck the surging breakers. The *Bonham* was anchored just off the *Lee*'s port quarter, close aboard, and the *Tuscarora*, which had arrived and assumed command of operations, was notified of the *Bonham*'s failure to affect the rescue. By this time, two small motor surfboats had arrived from the Wood End and Manomet Point Stations. Boatswain Brown recommended to the senior officer present aboard the *Tuscarora* that these two motor surfboats—less unwieldy—be utilized in removing the passengers. Accordingly, the *Tuscarora* advised the *Bonham* that the two surfboats, together with the CG 176 and the patrol boat *Active*, were under its command.

Boatswain Brown directed the two surfboats to proceed at once to the *Lee*'s side, start removing the passengers, and bring them in boatloads to the *Bonham* and the cutter *Redwing*, which had since reported on the scene.

After several attempts, a surfboat managed to get under the *Lee*'s rail, and at 10:20 A.M., the first boat load—twenty-seven women and children—was taken from the stranded vessel to the *Bonham*. At 11 A.M., the second surfboat came alongside the *Bonham* with twenty-two more passengers. Boatswain Brown ordered the CG 176 to go around under the *Lee*'s rail, which the patrol boat accomplished, removing sixty-four passengers. Brown decided that the transfer of the passengers from the 176 to the *Bonham* would be too dangerous and accordingly directed the patrol boat to proceed to Plymouth to discharge them.

The surfboats, however, again drew alongside the *Bonham*, and this time, twenty-three passengers were taken aboard. Word was received by Boatswain Brown that a pulling boat from the Manomet station had capsized close to the beach and that the men who manned it were in desperate need of assistance.

The two surfboats were dispatched to the scene, where two Coast Guardsmen, almost unconscious and half frozen, were found clinging to the bottom of their capsized craft. The Wood End motor lifeboat, under the command of Boatswain Gracie, ran into the tremendous surf among the rocks and, with careful handling, managed to get close enough to pick up the two men. At all times, there was the possibility of the surfboat being thrown upon the sharp rocks, but the rescue was effected without accident.

From the survivors, Wood, Proctor, and a civilian named Douglas, who had volunteered to go in the boat to make a full crew, the story was learned in detail: The Manomet Point Station crew, under the command of Boatswain's Mate First Class W. H. Cashman, had attempted again and again throughout the night to launch their boat through pounding surf in an effort to get to the stranded *Robert E. Lee*. The passengers were in immediate need of assistance, and it was the Coast Guardsmen's job to affect a rescue. Six times the slashing surf hurled their staunch little craft back onto the beach, but the heroes who maimed it did not quit.

In the gray dawn, another attempt was made but with similar results. By this time, being worn out with the terrific exertion of battling the surf, the men were obliged to pause and rest.

Around 10 A.M., they again tried to get their lifeboat into open water. This time, they pulled desperately through the thundering surf, up, over, and through the breakers until clear water was under them. They made their way to the side of the stranded steamer and rendered valuable service in handling lines to and from the rescue vessels. After an hour alongside the *Robert E. Lee*, the lifeboat began its hazardous return trip.

As it approached the surf, three huge breakers came up astern and broke upon it. Under Cashman's able direction, the craft rode out two of the breakers, but the third roared over them, picking the boat up and smashing it down into the water, bottom side up. Instantly, the Coast Guardsmen were benumbed by the intense cold. One of them, Stark, was struck on the head by an oar and knocked unconscious. The civilian, Douglas, and a Coast Guardsman named Wood, went to Stark's assistance and managed to get him to the overturned boat. Together, they pushed him up on the keel and fastened his hands around it, hoping that they would freeze in that position, thereby keeping him from being swept away by the waves.

Then, Douglas and Wood clamped their own hands around the keel and began the long wait for rescue. The terrible cold ate into their bones; consciousness came and went. Cashman, the man in charge of the boat, had been thrown a considerable distance inshore, and his companions saw him fighting desperately for his life. He struggled through the surf, now making a yard, now losing two. Unable to assist him because of their own benumbed limbs, Douglas and Wood watched helplessly. Finally, Cashman disappeared forever in the maddened seas.

The other man, Griswold, was not seen after the boat capsized. Stark and weakening fast, he was washed from the keel of the boat while Wood and Douglas were unconscious and lost his life.

In the *Robert E. Lee* case, the Coast Guard rescued 323 passengers and crew and, in doing so, lost three men—Cashman, Griswold, and Stark. Three men who added glory to the Coast Guard. Three men who would have disliked being called heroes; they did only what they thought they were supposed to do—die, if necessary—to save others.

That spirit epitomizes the men of the Coast Guard—stout-hearted men who live by the sea and who, all too often, die by the sea. On watch in hundreds of lonely stations, at sea in pitching ships, or tramping miles of desolated beach, their everyday deeds of valor and self-sacrifice are often unsung. That they give their lives in the pursuit of their duty is part of the tradition of the United States Coast Guard.

SEMPER PARATUS! Always Ready—the Coast Guard!

Chapter XI

FLOODS AND HURRICANES

The flood along the Ohio and Mississippi Valleys in January and February 1937 has been characterized by the nation's welfare agencies as the worst peacetime disaster in the history of the country. While the loss of life was unbelievably small—slightly in excess of three hundred persons having met death as a direct result—property damage was estimated at hundreds of millions of dollars.

Usually, the floods along these valleys happen during the latter part of March or early April, the result of a heavy thaw after a long and severe winter in the lands surrounding the headwaters of the Ohio and Mississippi Rivers. The flood of 1937, however, began soon after the middle of January, in what has been observed as an unusually mild winter, with little snow, mild temperatures, and heavy rains. Being in the form of rain, this precipitation was not held in check by freezing weather; thus, the rush of water started down into the large river basins, with the consequent overflow of the banks. Rapidly, the waters rose, sweeping out over millions of square miles of the richest land in the world, inundating prosperous cities, paralyzing communications, and striking with such devastating suddenness that thousands of persons were trapped in the flooded areas.

Swiftly, the Coast Guard swung into action. On January 19, the Chicago Division commander was directed by headquarters to send six boats to Evansville, Indiana, and there report to the regional director of the Red Cross. This was merely the beginning. By the time the flood waters reached their height, the Coast Guard had in the flood area over 350 boats, fifteen planes, twelve communications trucks, and nearly 2,000 men.

From January 19 to March 11, 1937, the Coast Guard saved from immediate peril nearly 1,000 persons, evacuated 67,000 refugees, transported 16,000 relief workers, doctors, and nurses, saved and removed to places of safety nearly 2,000 livestock, and recovered six bodies. In addition to this work, the service carried several tons of mail, towed disabled boats and floating houses to safety, helped restore and maintain telephone and telegraph communications, established patrols to aid in the prevention of looting, carried food and medicines to the destitute, removed the sick and injured, helped bury the dead, and even had the forethought, as the flood waters receded, to drain the water from the radiators of tractors and automobiles to prevent damage from freezing.

Besides the rare efficiency and devotion to duty exhibited by the officer and men who made up this great relief force, two additional traits emerged to elicit praise from all who came in contact with the Coast Guard relief work—initiative and zeal. At the beginning of the flood, supervision by commissioned officers was impossible as there was no time to organize and lay down a plan of operation. Warrant officers and enlisted men were sent into the flood area in boats with no further instructions than to do what was needed.

Telephone and telegraph communications had been suspended because flood waters, often rushing along at a speed of 15 mph, had swept away the lines; the men had to be relied upon to reach the local authorities, local pilots, and others who could furnish information regarding those in distress. The Coast Guard boats proceeded under the most difficult and hazardous conditions, navigating among floating structures, telephone poles, trees, and under heavily charged electric wires, under low bridges, over railroad tracks, usually in rain, sleet, snow, fog, and often in pitch darkness. Somehow, these boats reached those in need, and the Coast Guardsmen maintained tender care over the sick, the aged, and the children; they even divested themselves of their own clothing so that their charges might be more comfortable.

The true worth of the Coast Guardsmen in this great national emergency was fully realized when, after three or four weeks in remote areas helping those in trouble, they began to trickle back, tired, hungry, often ill, but carrying with them letters of appreciation from dozens of communities, praising the work these men had done, work that had been spontaneous and without direction, work done with the single thought to serve and to serve well. These men were cheerful and obedient at all times, faithful to traditions and ideals that have been the Coast Guard's foundation down through the years; they were ready to proceed again and again into the desolated areas so seldom visited by sailors.

As scouting areas could not be reached immediately by boat, Coast Guard aviation rendered invaluable service in transporting doctors, nurses, and serum. Landings and takeoffs were made under the most hazardous conditions, with the ever-present possibility of colliding with submerged objects. Information relative to the safety of those persons residing in areas where levees were reported to be weak was gathered by the planes; as the center of the disaster moved down the great river valleys, personnel were transported from one point to another.

The full efficiency of the Coast Guard communication system was strikingly brought to light when the communications trucks, fully equipped with the most modern radio installations, were sent into the stricken areas, traveling far and wide but reaching their designated posts, through swamps, over flooded roads, and reporting to the central communication headquarters. Besides providing listening posts for the relief agencies, these trucks afforded a rapid and efficient means of communication with those cities and communities long since cut off from the outside world by the flood waters. Working night and day, the operators kept relief headquarters advised of the conditions and needs of those persons in their immediate radius.

Instances of unusual devotion to duty and personal heroism in carrying on the relief work came from all sides; everyday reports told how some Coast Guardsman, acting on his own initiative, had risked his life in order that others might be served. Such were the actions of Willis P. Wills, boatswain's mate first class—regularly assigned to the Cape Hatteras Coast Guard station on the East Coast but now serving with the relief forces—in performing courageous and unselfish work at the risk of his own life at Huntington, West Virginia, on the morning of January 30, when the flood was at its height.

Information had been received by the city authorities at Huntington that a large gasoline tank in the center of the city was leaking, its inflammable contents spreading over the water for a considerable distance. Here was a dangerous situation. A carelessly thrown match or a spark from an electric wire would result in a conflagration with which the crippled city would be unable to cope. Something had to be done.

It was Coast Guardsman Wills who volunteered to take firefighting equipment to the tank. Local authorities impressed upon him the dangers: A flare from the exhaust of the surfboat might cause an instant explosion, and if that happened, no one could help him.

But Wills' mind was made up. The gasoline was spreading downstream to other buildings; sooner or later, it would meet an open flame. Then—

Loading a surfboat with firefighting equipment, Wills started slowly through the dangerous waters, maneuvering carefully so as not to strike submerged wires and metal. The men in the buildings nearby watched him breathlessly, for they realized that if an explosion occurred through the contact of the craft's exhaust with the gasoline, all of them would be trapped.

Wills reached that tank safely, and soon he had taken all the precautions necessary to prevent a fire.

On the same day, January 30, another Coast Guardsman, Surfman E. M. Gray, of the False Cape Station, at False Cape, Virginia, was singled out by high Army officials as another outstanding hero of Huntington's flood crisis. With his crew of a Coast Guard surfboat, Surfman Gray rescued eighty-nine persons from immediate peril in one afternoon.

The Army command in charge of operations in the vicinity of Huntington was lavish in its praise of Gray's work: "Other men have, no doubt, done valiant service, but this man's work is about the best that has been reported up to this time. The interesting fact about the record was the efficiency with which he went to work on his own with only general instructions. He was instructed to go to the eastern part of the city and start rescue work, and his showing of eighty-nine persons saved in an afternoon seems to indicate that he organized his work with exemplary efficiency. We think that it is an outstanding example of fine discipline and level-headed action."

At Mound City, Illinois, Coast Guardsmen evacuated the town's total population, which numbered nearly 2,000, and then learned that about fifty of the community's citizens had returned and were holding on for dear life in the face of the still-climbing waters. Cruising among the partially submerged homes, the Coast Guardsmen found several women and children living in attics, hoping desperately that they would not be compelled to give up their homes. The Coast Guardsmen removed these people, however, after stressing the danger of the houses collapsing from the onslaught of the water.

In one house, they found an old nanny standing guard over three children whose parents had disappeared. The old woman had been afraid to leave for fear that the children's parents would not be able to find them. The house was beginning to fall apart, and after cajoling failed, the Coast Guardsmen forcibly removed her and the three children and took them to a packet steamer standing by to transport refugees.

A bearded old man refused to leave until his cow, Old Betsy, was first placed aboard a rescue craft. Under one arm, he carried a cage with two canaries; under the other, a small white dog. "Yessir, hit's a big flood," he told Coast Guardsmen. "Reckon hit's the biggest flood I ever did see. I'd been away from here two days ago if it hadn't been for Old Betsy. I couldn't go without her. Thar she is, over there, moored to that lumber pile. Ain't she a beaut?"

Early in the morning of January 25, an Ocean City motor surfboat, in charge of Boatswain's Mate First Class (Life-saving) William F. Burton, was cruising up Vine Street in Cincinnati, Ohio, cutting electric wires, when he heard screams in the darkness. Investigation showed three men in the river, two of them clinging to a capsized rowboat, while the other was clinging to a bridge, which was already tottering with the water's sweeping force.

The two men were rescued promptly from the rowboat, but the surfboat could not approach close enough to the bridge to remove the third man. Surfman Samuel C. Mitchell promptly secured a line around his waist and, crawling hand over hand along the girders of the bridge, managed to reach the unfortunate man. By this time, the bridge was sinking lower and lower into the water and was tilting slowly over on one side. Mitchell worked rapidly. Securing the line around the man's waist, he signaled for the men on the surfboat to haul away. Then he started back the way he had come, forced to crawl through icy water that now swept over the girders.

At last, he reached the surfboat, but the work of this tireless Coast Guard crew was not done. They continued their rescue work for four days without sleep. Keeping to their feet in the rain and sleet and snow that was whipping around Cincinnati, they added to the glorious record that the entire Coast Guard was making in this, the greatest of peacetime disasters.

On September 18, 1938, ships in the South Atlantic flashed warnings to the United States Weather Bureau that a hurricane was brewing in the south. Subsequent reports plotted the path of the disturbance northward toward the Florida Keys at 17 mph, but it abruptly shifted its course, glanced off the Florida coast, and proceeded past the Carolinas. Experts predicted that it would curve out to sea and expend its force there without damage. By Wednesday, September 21, it had passed Cape Hatteras, and storm warnings were hoisted along the coastline all the way to Eastport, Maine. No further news was received until wind of terrific force began to beat the coast near Atlantic City, New Jersey.

Meteorologists along the New England seaboard observed an alarming fact. The hurricane had covered a distance of 600 miles in twelve hours, one of the fastest movements ever reported. Somewhere along the coast, it had suddenly burst with tremendous velocity and was now sweeping toward the New England states.

Like lightning, disaster struck the entire coastline. Whole settlements were blasted out of existence, swept away by tidal waves brought in by the hurricane. Coast Guardsmen worked desperately, almost helplessly, to save those who were trapped by the raging wind and sea. At New London, Connecticut, the full force of the storm struck at 3:30 P.M. on September 21, and when the wind cups of the Naval anemometer blew away, the hurricane was blowing at a speed of 98 mph. Its actual force has been estimated at from 125 miles to 175 miles an hour. Within four hours after the hurricane struck, the winds began to slacken; but already whole sections of New London lay in ruins, and one-quarter of the business section was in flames.

Early in the morning of September 21, Chief Boatswain's Mate Ulric F. Engman, U.S. Coast Guard, attached to Base Four in New London, was departing on leave when he noticed the increasing winds and lowering darkness hovering over the city. Returning to New London, he resolved to report back to the base in order to be available for duty in case he should be needed. By this time, the forerunning winds of the hurricane, together with heavy rain, made it impossible for Engman to reach the base. Instead, he reported to the New London Police Station and offered his services to that department. Within a short time, the full force of the hurricane was roaring upon the city.

Shortly after Engman's arrival at the police station, a report was received that a man was drowning in Winthrop Cove. Police Sergeant Courtney, who was on duty at that time, directed Engman to proceed to the Cove and attempt to rescue the drowning man. Upon Engman's arrival there, he found about two hundred persons watching an old man struggling for life aboard a small boat, which was dragging mooring about two hundred yards from the shore. Heavy seas were washing over the boat, and it appeared that the small craft was in imminent danger of going to pieces. The water, filled with floating debris, was about seven feet above normal tide level.

Engman knew he must act at once, for the old man was having a desperate time trying to keep the craft's bow headed into the threatening seas. Finding a line, the Coast Guardsman fastened it around his waist. Then, after ordering the spectators to pay the line out carefully and to try to keep it free of debris, he jumped into the water.

Engman found the going very hard; he had to be constantly alert to keep floating logs and wreckage from crushing him. He fought on through heavy seas, now making a yard, now falling back a foot or so. Body bruised, muscles aching, half-strangled by high, buffeting waves, Engman stuck at it while the hundreds on shore watched in breathless amazement. They were seeing a man at his best, fighting a fight that could have only one ending if he lost—death. This was a splendid exhibition of nerve and physical ability, one man's struggle in the course of duty to save another's life.

Engman was suddenly confronted with a new danger. He was compelled to swim around a warehouse building, already tottering, that had been swept out into the Cove. Fighting desperately, his strength waning, the Coast Guardsmen made his way around this floating menace and came at last to the bot. Grasping the sides of the craft, he rested for a moment while the waves beat at him unmercifully.

At this point, a civilian, later identified as a Mr. Stinka, jumped in and came to Engman's assistance. Together, the two men managed to fasten Engman's line to the bow of the boat. Then the tremendous job of unshipping the anchor chain from the anchor to allow the boat to float free was begun. It was heartbreaking work. The seas broke over them, pummeling them mercilessly, threatening every moment to sweep them away to death. Finally, however, the anchor line was cut, and the spectators on shore pulled the boat closer to land.

With Stinka's assistance, the old man—Carl Petersen—was lowered to Engman's back, and the Coast Guardsman started the return to shore. Sixty-four years old and disabled, Petersen's dead weight gravely taxed Engman's already depleted strength; somehow, the Coast Guardsman won through to safety, where willing hands helped him and the old man. Suffering from shock and exposure, Petersen was taken to the police station for first aid and later sent home.

As the force of the hurricane swept further north, other Coast Guardsmen swung into the rescue and relief work. At Penzance Point, Woods Hole, Massachusetts, the Coast Guard cutter *General Greene* dispatched a rescue party to aid distressed families along the shore. Among this party were Radioman Third Class John A. Steadman and Seaman First Class Charles G. Starling. The Coast Guardsmen proceeded to Penzance Point, where, it had been learned, two men were trapped on a telephone pole.

Arriving at the scene, Starling fastened a line around his waist and attempted to wade through the surging waters to the aid of the men on the pole. Fighting desperately, he made three attempts to get to them, but each time he was swept off his feet by the force of the roaring water.

Exhausted, sore in every muscle, the gallant Coast Guard seaman was about to make a fourth attempt when one of the men lost his hold on the pole and tumbled into the water to be swept instantly to death.

Starling tried to reach the remaining man from another angle, working with the water instead of against it. Wading up to his armpits, falling and stumbling but managing in some manner to keep going, he finally reached the man's side. Then, securing the rope around himself and the man, he gave the signal to be pulled into high land. Starling was exhausted when, at last, he returned to the *General Greene*. The rest of the rescue party, however, were able to keep on with their work.

Meanwhile, Radioman Steadman joined a woman in a search for her twelve-year-old son. After a considerable time, the youth was found on a small island, over which breakers were roaring in rapid succession. Fighting his way to the boy's side, Steadman carried him to high land, where he watched with a pleased smile as the mother happily embraced her son. But there was more work to be done. Retracing his steps through the swirling water, he next gave aid to a stranded couple who wished to go to the Breakers Hotel because their home was not habitable.

The tremendous struggle Steadman had made in rescuing the boy began to tell on him. A towering wall of dirty water pounded down upon him, knocking him from his feet and washing him against a submerged hedge. Here, he held on, desperately trying to marshal his failing strength, while the wind roared around him as if in mockery.

A second wall of water crashed down upon the Coast Guardsman, submerging him and sweeping him away from the hedge. Reaching out in a despairing gesture, his hand grasped a guy wire, and he tried to hold on against the terrific force of the current.

People on the high land saw him raise his head in a last, brave gesture. Then, suddenly, he vanished.

Steadman's body was recovered a few days later, and the Coast Guard mourned the death of a man who had brought undying glory to the service he had loved. Steadman was posthumously awarded the Gold Life-Saving Medal for his work in rescuing the twelve-year-old boy. Starling received the same tribute for saving the man on the telephone pole.

In every major national disaster, individual cases of heroism are brought to light within the service. Often, all too often, names like Steadman and others are added to that long list of men who have died to keep traditions alive.

During the peak of the hurricane, Chief Boatswain's Mate (L) Alfred Volton, Officer-in-Charge of Cuttyhunk Coast Guard Station, Cuttyhunk, Massachusetts, assisted by Surfmen Norman P. Cuppels, James A. Yates, and Allan L. Potter, rendered aid to twenty-four distressed people.

At 2 P.M., word was received at the Coast Guard station that two men stranded on Copicut Neck were in desperate need of assistance. Manning a motor lifeboat with a dory in tow, Cuppels and Yates set out immediately from the station. Reaching Copicut Neck, they rescued the two men and started the long fight back to the mainland, landing the men aboard their yacht *Pelorus*, which was anchored in Cuttyhunk Pond.

After reaching the yacht, the two men, in company with another man, decided to return to Copicut Neck and attempt to salvage a dory and an outboard motor that they had previously beached there. Reaching the place safely, the three managed to get the dory away from the beach and were heading for the yacht when the heavy wind filled their craft, and they were thrown into the sea.

These operations had been watched by the Coast Guardsmen, and by using the station dory, Volton, Cuppels, and Yates pulled the men from the sea and landed them on Copicut Neck. After landing, the Coast Guardsmen abandoned the dory and, walking over to the beach, boarded the lifeboat at the city dock.

About the middle of the afternoon, ten men who had been working on the construction of the Cuttyhunk Station and who had taken refuge aboard a pile driver out in the harbor signaled that the pile driver was in danger of capsizing and that they wished to be taken off. These men were removed in successive trips by Volton, Cuppels, Yates, and Potter to a place of safety. Scarcely had these operations been completed when word was received that five fishermen on the town dock were in immediate danger of being swept away by the increasing seas.

The work of rescuing the five men called for seamanship unparalleled in small boat handling. Heavy seas were sweeping over the dock, with the ever-present possibility of crashing the lifeboat against the heavy timbers. Volton made several attempts to reach the men, each time being forced by the waves to sheer off.

Finally, the Coast Guardsman maneuvered his boat up to the side of the dock, and while the rest of the crew fended it off, the five fishermen leaped across into the lifeboat. They were landed on safe ground, and the lifeboat started on another trip across the harbor, where several boats were lying at anchor with distress flags whipping the hurricane.

Removing two men from these boats, the lifeboat scouted around in the harbor, searching for other persons in need of assistance. Sure enough, at about 5:30 P.M., two men in a skiff were seen trying to run a line to a fishing boat. The skiff capsized, throwing the two men into the sea. The Coast Guard boat went immediately to their assistance, picked them up, and landed them safely.

During the entire operation in the harbor, the Coast Guard boat was menaced in its rescue work by such debris as fish nets, lobster pots, buoys, and other objects which might easily have meant disaster.

During the hurricane, which raged unchecked for a number of hours over the New England states, untold cases of heroism were performed by the men who make up the United States Coast Guard. It is impossible to capture in one single chapter the unselfishness, the self-sacrificing courage these men exhibited in the hours of unparalleled terror during which the hurricane roared over them. Some of the Coast Guardsmen saw their own homes swept away, knew not what had happened to their families, yet they kept at their tasks, saving other people, working with ready hands at the jobs they knew must be done. The Coast Guardsmen along the New England coast showed the world what it means to be a Coast Guardsman in a time of national stress; they proved conclusively that the Coast Guard lives up to its motto, SEMPER PARATUS— ALWAYS READY!

Chapter XII

THE COAST GUARD AND
THE FUTURE

The Coast Guard is moving forward, step by step, with the progress of the United States. More and more obligations and duties are being imposed upon the service by the consolidation of government agencies and the formation of defense plans designed specifically to protect both North and South America from assaults by foreign nation aggressors.

The placing of the Maritime Service under the direct supervision of the Coast Guard adds another solemn responsibility to the ever-growing scope of influence of the Coast Guard. The Merchant Marine Act of 1936, as recently amended, provides that the United States Maritime Service, a voluntary organization of the licensed and unlicensed personnel of the Merchant Marine, shall be administered by the Coast Guard in order that a trained and efficient merchant marine shall be established by providing an adequate training system, and contributing benefits for seamen of good character and ability who serve aboard vessels of the Great Lakes and the high seas.

For 150 years prior to this legislation, there had been no determined effort on the part of the federal government to lay down the foundations of a Merchant Marine that would match growth with the nation's prosperity. Until comparatively recent years, there appeared no need for such governmental activity. From the Republic's earliest days until the 1860s, American ships could be found on all oceans, and a significant part of the growth of the country can be attributed to American merchant ships and the men who sailed them. In the early days of the American Republic, ocean commerce was of prime importance to its citizens since all of the original states lay along the Atlantic Seaboard and the vast majority of freight was carried by water.

However, as the United States gradually expanded inland and became absorbed in successive eras of agriculture and industry, the importance of the Merchant Marine was lost to the sight of the American people. The number of American ships dwindled from the oceans until over 75 percent of this country's commerce was carried in foreign ships. It is a matter of painful history that Dewey, when he was ordered to Manila at the outset of the Spanish-American War, was compelled to charter foreign ships to carry coal and supplies for his fleet.

Apparently, even this did not bring home to the nation the realization that the American Merchant Marine was practically non-existent, thus exposing a glaring weakness in the national defense system. In 1917, after war with Germany had been declared and the problem of transporting millions of men and tons of supplies to France presented itself, the United States began a hurried and costly attempt to create, on the spur of the moment, as it were, a Merchant Marine that would meet the demands that war had placed upon it.

The result was a mass production of ships, all kinds of ships, steel and wooden. At the end of the war, the United States had afloat a vast fleet of cargo vessels, with no trained and experienced personnel to man them. Thus, it came home to the American people that a Merchant Marine cannot be built upon the spur of the moment nor within the space of two years. The foundations must be put down carefully, the importance of discipline must be stressed, and men must be trained thoroughly in the things they will be called upon to do at sea.

The United States Coast Guard has undertaken that job. It is a big task, but the service has the background and the experience to carry it through to a successful completion. There are no better sailors in the world today than the Coast Guardsmen, for it is their business to go to sea in small boats, which require the most expert handling in heavy seas.

Coast Guard Commander William N. Derby is charged directly with the training of both officers and seamen of the United States Merchant Marine. His office works in the closest cooperation with the Maritime

Commission, two members of which are high-ranking Naval officers—
Rear Admiral Emory S. Land and Rear Admiral Henry T. Wiley. It is
the work of these two officers to bring the Merchant Marine closer to
the Navy, thus coordinating the actions of the two branches against the
possibility of war.

Three Maritime Service Training Stations, under the direct supervision of the Coast Guard, with Coast Guard instructors, have been
established; two of these stations are on the East Coast, one at Hoffman Island, New York Harbor, for the training of unlicensed personnel,
another at New London, Connecticut, for licensed personnel. These two
stations serve seamen from the Atlantic ports, the Great Lakes, and the
Gulf, while the training station on the West Coast, at Government Island, Oakland, California, receives both licensed and unlicensed personnel from the Pacific Coast ports.

Enrollment in the Maritime Service is limited to citizens over twenty-
one years of age who have served at least two years on merchant vessels
of the United States of five hundred gross tons or more, operating on
the Great Lakes or any ocean, seven months of such service being within
two years of the date of application for enrollment. An applicant must be
able to pass a strenuous physical, mental, and moral test in order to prove
his fitness for service on a seagoing vessel. Licensed personnel who have
served at least seven months as officers under the requirements of their
licenses in charge of a watch are eligible for enrollment as ensigns in the
Maritime Service. Unlicensed personnel are eligible for enrollment under
their different branches of service.

These enrollments are made for a probationary period of three
months, the enrollees being furnished transportation by the service to
the training stations. Quarters, meals, and a uniform outfit are supplied
by the government without charge. The enrollees are instructed in all
subjects relating to the branch of service under which they serve; the
courses of instruction are under the direct supervision of Coast Guard
officers and men. At the end of two months, those enrollees who have
proven themselves fully qualified are advanced in grade and pay.

At the completion of his three-month period of probationary training, the enrollee is given transportation from the training station to the

place of his enrollment. If his qualifications and conduct are satisfactory, he is offered enrollment in a regular status in the Maritime Service before he leaves the training station, with the understanding that, upon his release from active duty, the enrollee will serve annually for eight months on a seagoing merchant vessel and one month in the Maritime Service.

In developing a training course to fit the need of the Merchant Marine, close attention was given to the wide variation in the abilities and requirements of the enrollees, and every effort was made to provide each man with a course of training individually suited to his particular branch of service. At the Fort Trumbull Training Station in New London, Connecticut, which is open to licensed personnel only, successive classes of one hundred Merchant Marine officers each are held, and in addition to the training facilities available at other stations, the laboratory equipment of the United States Coast Guard Academy is utilized for instructing enrollees.

The Maritime Service employs four training ships, two of which are square-rigged vessels, the *Joseph Conrad* and the *Tusitala*, both based at Hoffman Island, New York. The Coast Guard cutter *Northland* has been assigned as a training ship for the Government Island Station in Oakland, California. The Maritime Service Training ship *American Seamen*— manned by a Coast Guard crew and commanded by Coast Guard Lieutenant Commander Charles Etzweiler—is specially fitted as a training vessel. It has accommodations for two hundred enrollees and is equipped with four complete shops, various types of lifeboats with their attendant gear, and all navigational equipment of the most modern character.

After an enrollee in the Maritime Service has finished his period of training and leaves the station, a number of correspondence courses administered by the Coast Guard Institute in New London are available at no cost and thoroughly cover all phases of seamanship.

The American Merchant Marine is on the upswing again. A program has been undertaken wherein five hundred merchant vessels are scheduled to be constructed within ten years! These ships are to be the best that science and the magic of engineering skill can produce; they will have incorporated within their construction several national defense features which will make them truly remarkable vessels. Four of a series of twelve high-speed tankers have already been launched, and they and their sister ships have been ordered by the Standard Oil Company of New Jersey. Trials have shown that these ships can develop over four times as much horsepower as an ordinary tanker, being capable of attaining a speed of 20 knots with a full load of 150,000 barrels of oil.

At Newport News, Virginia, the biggest ship ever to be laid down in an American shipyard is under construction and, in spring, will lead the parade of the new American Merchant Marine. Named the *America*, she will be one of the finest and safest ships afloat.

This reconstruction of the Merchant Marine will present a real demand upon the protective services of the United States Coast Guard, and the manning of such a huge commercial fleet will tax the Maritime Schools to the limit. Already, plans are being worked out with the intention of doubling the normal output of 4,000 men trained annually by the Coast Guard. The ultimate aim of this comprehensive maritime program is to have a Merchant Marine second to none, fully trained and ready to take his place in the parade of American progress.

The Coast Guard moves forward!

In July 1939, the Coast Guard, the second oldest bureau of the government, absorbed the Bureau of Lighthouses in accordance with the provisions set forth in President Roosevelt's reorganization plan. Oldest of all government agencies, the Bureau of Lighthouses thus passes as a separate unit of the government; with the consequent induction of the lighthouse personnel into the military establishment of the Coast Guard, it becomes a part of the Coast Guard. The general consolidation plan for the two services is to take these men into the Coast Guard, as far as is practicable, in ranks, grades, and ratings analogous to the pay and responsibilities to which they had been accustomed in the Lighthouse Service. It has been estimated that the Coast Guard personnel will reach a total of 18,000 men after the absorption has been completed. In order to establish a single promotion list for both services, such personnel entitled by their former rank in the Lighthouse Service to commissions will be inducted into the Coast Guard line.

To accomplish this great consolidation, a complete reorganization of the Coast Guard and the Lighthouse Service has been considered.

In fact, a new system of Coast Guard districts (replacing the old Coast Guard divisions and districts and the lighthouse districts) has been installed, with the former division commanders taking over the rank of district commanders and being charged with the responsibility of both the Coast Guard activities and those of the former Lighthouse Service. These commanders will be assisted by a Coast Guard officer, who, before his commissioning, had served as a lighthouse district superintendent.

The Lighthouse Service brings to the Coast Guard traditions and stories steeped in the essence of American history. The first lighthouse to be erected in this country, probably the first on the American continent, was built by the province of Massachusetts in 1716 on Little Brewster Island at the entrance to Boston Harbor. This work was accomplished at the cost of approximately $10,000, and an old mezzotint of 1729 pictures it as a tall and imposing structure of gray stone. The light was supported by a tax of one penny per ton on all incoming and outgoing ships except coasters. These vessels paid two pence each upon clearance, and all fishing vessels were required to pay five shillings each year.

Here, at this old station, the first fog signal to be established on this continent was put into service. In 1719, the keeper of the lighthouse requested the general court to place a large gun on the island for the purpose of answering ships in the fog. This old gun, dated 1700, may still be seen at the Boston Light Station.

The history of the Lighthouse Service as an administrative agency of the national government is practically coincident with that of the nation itself. The first president, George Washington, took a personal interest in the lighthouses. One of his first official acts upon becoming president was to write a letter to the keeper of the lighthouse at Sandy Hook, directing him to keep the light burning until Congress should provide funds for its upkeep. The Lighthouse Service formally became an administrative part of the government by an act approved on August 7, 1789—the ninth law to be passed by Congress, the first provision to be made for any public work. The Lighthouse Service was created one year before the Coast Guard.

After a long delay, the original states finally deeded the various lighthouses, which they had built and operated, to the federal government—with the understanding that the United States would be responsible for their upkeep and the salaries of the keepers. An old letter from the president's secretary, Tobias Lear, to the Secretary of the Treasury shows that the salaries paid to the keepers were small even for those days, ranging from $120 to $266.66 per year.

The Lighthouse Service, with the location of the lighthouses and lightships in such strategic positions and the continuous patrol of the coastal waters by the tenders, will prove a valuable adjunct to the Coast Guard in its life-saving duties. It brings with it an enviable record in rescue work, and the fact that the service has been primarily concerned with other duties makes the risks and hardships taken in these rescue cases all the more commendable.

In 1916 alone, 165 cases of life-saving were reported, and in July of the same year, the tender *Cypress* did yeoman work in removing twenty-two persons from a sinking collier during the height of a hurricane, under circumstances requiring the highest courage and seamanship.

On board the *Diamond Shoals* Lightvessel, there hangs a framed letter from President Roosevelt, commending the lightship's crew for the manner in which they handled their duties during the hurricane of September 15-16, 1933: "I have read with keen satisfaction the report of the heroic work done by the officers and crew of the *Diamond Shoals* Lightvessel during the hurricane of September 15-16. I am fully appreciative of the exceptional character of the service performed in saving this vessel and in the protection of the shipping along the coast; and I wish you would convey to them my personal commendation for the manner in which they performed their dangerous duties during this storm."

This hurricane reached an estimated velocity of 120 mph, its center touching the Atlantic Coast at Cape Hatteras and passing directly over the lightship's station on its journey seaward. The following report of the master of the *Diamond Shoal* Lightship, C. C. Austin, gives a keen insight into the terrible experiences suffered by the men aboard the ship:

On the morning of the fifteenth the weather showed indication of a hurricane. At 8 A.M., wind east between forty and forty-five miles per hour, increasing, barometer falling. I got the engine underway and began to work ahead slow. From noon to 4 P.M. wind east-northeast between fifty and sixty miles per hour, increasing, barometer falling. Seas getting rough and washing ship badly.

At about 2 P.M., station buoy sighted for the last time as the weather was thick with rain and spray. I judge the ship began to drag anchor at about 4 P.M., wind increasing to about seventy miles per hour. I began to increase the speed of the engine from forty to sixty revolutions per minute. From 8 P.M. to midnight, wind east-northeast, between seventy and eighty-five per hour, barometer falling. Seas were getting mountainous high and washing the ship terribly. Engine speed increased to ninety revolutions per minute.

September 16, between midnight and 1 A.M., ship went into breakers on southwest point of Outer Diamond Shoals (having dragged the fifty-five-hundred- pound anchor and twenty-four thousand pounds of chain the five miles from her station). Wind about one hundred and twenty miles per hour. The first breaker which came aboard broke an air port in the pilot house which struck me (master) in the face and around the neck and on arm, cutting face and neck badly. This same breaker carried away one ventilator close to the pilot house. Mate S. F. Dowdy tried to get a stopper in the hole in the deck, and washed against a davit and broke some ribs. He was almost washed overboard. From 4 to 5 A.M. wind decreasing to about fifty miles per hour, barometer falling to 28.19 (lowest point).

We laid in the breakers from 12 midnight until 6:30 A.M., breakers coming aboard, breaking up everything on upper deck, washing boats, ventilators, awning stretchers away, bending awning stanchions inboard. Taking water in around umbrella of smokestack and through ventilators to such an extent that the water was rising at times above fire-room floor with all pumps going, and every means we had to keep the water out of the ship.

At 5:30 A.M., day began to break, so I could see the conditions outside. I could see an opening about south-southwest from the ship that looked like a chance to get away. Breakers coming over at intervals and I decided that it was the only chance out. I told the mate to get ready to slip the mooring, as we had to get out of that place, for when the wind comes from the west it would carry her into the breakers and finish her up. I slipped the mooring at 6:30 A.M. and got the ship outside the breakers, at about 7:15 A.M. being in the center of the hurricane. I had just got the ship clear of the breakers when the wind struck from the west at about ninety miles per hour. I ran the ship southeast until I was sure I was all clear and then ran northeast thinking the hurricane would pass. I ran this course for a while and it did not get any better. I considered it was moving very slow (the barometer was rising very fast) so I changed my course to the south and ran this course until I ran out of the hurricane.

September 17, 5 A.M. Wind northwest, strong gale, but decreasing. At 6 A.M. I called the mate and told him to get the crew out and see if he could get the wireless antenna fixed up so that we could establish communication. (There had been no radio communication since Friday evening.) At about 9 A.M., I got radio compass bearings which put the ship approximately sixty miles east-northeast from Cape Hatteras Lighthouse; at 4 P.M. radio bearings placed ship about one hundred and ten miles east-southeast from Cape Henry. All the crew were at hand at all times and ready to do everything they could to help save the ship, both deck and engine force. During the storm one of the fusible plugs in the boiler blew. They let all steam from the boiler and opened up the furnace, went inside and took out the fusible plug that had blown and put in a new one, and closed the furnace and got steam on the boiler in the strength of the hurricane. I consider this a brave deed, and M. W. Lewis and J. J. Krass, firemen and A. D. Ameyette, seaman, are due all credit for accomplishing this job. I consider each and every man of the crew did all in his power, and through their bravery, energy and will power, we brought the ship through the hurricane and safely into port. The vessel I consider a most excellent seaworthy ship to come through such a severe hurri-cane with such comparatively slight damage as was sustained; so much water came aboard that at times there were three feet of water in the engine-room bilges.

These are the sort of men who are coming into the Coast Guard, bringing with them traditions and examples of unwavering devotion to one main ideal; to be of service to those who travel the sea. In this, the two services have never been far apart, and it is natural that they should unite for the further development of that ideal. While the Lighthouse Service has passed out of existence as a separate agency of the government, the exploits and glowing record of achievement which it has maintained steadfastly throughout 150 years of independent life will remain irrevocably its own and will tend to enhance the name, Lighthouse Service, in the hearts of the American people.

In May 1939, a bill to establish a Coast Guard Reserve, to be composed of owners of motorboats and yachts, was submitted to Congress. The purpose of the bill is to promote interest in the safety of life at sea and upon navigable waters, the promotion of efficiency in the operation of motorboats and yachts, and a wider knowledge of and better compliance with the laws and rules and regulations governing the operations and navigation of such craft and to facilitate certain operations of the Coast Guard.

The Coast Guard, as the nation's maritime police force, is charged with the duty of boarding the 300,000 motorboats and the 4,000 yachts operating in our waters to see that they comply with the navigation and other laws and also to assist them when they get into trouble.

Since 1935, many letters have been received by the Coast Guard requesting that a Reserve be established, but the actual plan had its birth in the Coast Guard Headquarters in Washington. It was put forth as a means of better law enforcement on a voluntary basis and the furthering of competency in motorboat operation and navigation.

The plan is to form a Coast Guard Reserve whereby motorboat and yacht owners may enroll in the Reserve under the rules and regulations drawn up by the Coast Guard after an examination of the owners and an inspection of the boats. Those who pass will receive a Coast Guard Reserve flag to fly and will be given a uniform insignia. Penalties for the unauthorized usage of the flag and insignia will be provided. The existence of such a Reserve and the flying of such a flag by those qualified will prove an incentive to others who are not as proficient in acquiring competency in the various factors relating to the operation and navigation of motorboats and yachts so that they may also have the privilege of flying the Coast Guard Reserve flag.

In 1938, there were more than 14,000 cases of assistance—largely incident to the operation of motorboats—handled by the Coast Guard. Some of these craft put to sea without the necessary life preservers, others without compliance with the law concerning the engines and seaworthiness of the boats. The purpose of the Coast Guard Reserve is to reduce the number of boats which get into difficulty because of the incompetence of the operators, lack of proper equipment, or lack of compliance with the law. The Coast Guard intends, by means of this Reserve, to disseminate among owners of motorboats, and particularly small fishermen, a knowledge of the regulations and laws applicable to motorboats so that they will not incur the many fines and inconveniences now imposed upon them.

There is nothing compulsory about the Coast Guard Reserve. It is purely a voluntary organization to increase the safety of life at sea and to better the compliance with the laws without recourse to punitive measures. It will be a matter of pride for the owners of the motorboats traveling the Chesapeake, the Potomac, or any other place, to be able to display the Coast Guard Reserve flag at the masthead. For this flag will indicate to the world that they have been examined and passed on, that they know the rules of the road, and that their vessels are seaworthy, properly equipped, and competently operated.

Last year, the fact that the Coast Guard patrolled nearly 500 regattas and marine parades in the various sections of the country imposed upon it a great burden in the handling of requests for marine supervision. The Coast Guard Reserve will afford to the service a broader field of boats to call on for the patrol of these regattas; by putting aboard a Coast Guard officer the Reserve boats, the effective range of the Coast Guard in these duties will be increased. The use of a Coast Guard officer on the Reserve boat will be, of course, with the permission of the owner of the boat since the Reserve is conceived wholly on a voluntary basis.

On this point, Admiral Waesche, the Commandant of the Coast Guard, said: "Being enrolled in this Coast Guard Reserve means the Coast Guard has placed its stamp of approval on that man as knowing the rules of the road, knowing how to operate a motorboat, and anyone else in that locality who does not know much about it can go aboard that boat and get information. So it is, in a measure, an enlarging of the facilities of the Coast Guard without any expense whatever to the federal government with the exception of the very slight cost for the administration of the act."

The bill provides for the expenses incurred while the Reserve boat is at the disposal of the Coast Guard to be borne by the Coast Guard, but no compensation is provided for the personal services of the operators. The Reserve bill further states that the members of the organization, solely by reason of their membership, are not vested with, nor can they exercise any right, privilege, power, or duty vested in or imposed upon the personnel of the Coast Guard.

www.ingramcontent.com/pod-product-compliance
Lightning Source LLC
Chambersburg PA
CBHW021145090426
42740CB00008B/951